Water Flow in Soils

Second Edition

BOOKS IN SOILS, PLANTS, AND THE ENVIRONMENT

Soil Biochemistry, Volume 1, edited by A. D. McLaren and G. H. Peterson

Soil Biochemistry, Volume 2, edited by A. D. McLaren and J. Skujins

Soil Biochemistry, Volume 3, edited by E. A. Paul and A. D. McLaren

Soil Biochemistry, Volume 4, edited by E. A. Paul and A. D. McLaren

Soil Biochemistry, Volume 5, edited by E. A. Paul and J. N. Ladd

Soil Biochemistry, Volume 6, edited by Jean-Marc Bollag and G. Stotzky

Soil Biochemistry, Volume 7, edited by G. Stotzky and Jean-Marc Bollag

Soil Biochemistry, Volume 8, edited by Jean-Marc Bollag and G. Stotzky

Soil Biochemistry, Volume 9, edited by G. Stotzky and Jean-Marc Bollag

Water Flow in Soils

Second Edition

Tsuyoshi Miyazaki

University of Tokyo
Tokyo, Japan

CRC Press
Taylor & Francis Group
Boca Raton London New York

CRC Press is an imprint of the
Taylor & Francis Group, an **informa** business

CRC Press
Taylor & Francis Group
6000 Broken Sound Parkway NW, Suite 300
Boca Raton, FL 33487-2742

First issued in paperback 2019

ISBN-13: 978-0-8247-5325-2 (hbk)
ISBN-13: 978-0-367-39243-7 (pbk)
Library of Congress Card Number 2005012832

Library of Congress Cataloging-in-Publication Data

Water flow in soils / edited by Tsuyoshi Miyazaki.-- 2nd ed.
      p. cm. -- (Books in soils, plants, and the environment ; v. 112)
  Includes bibliographical references and index.
  ISBN-13: 978-0-8247-5325-2 (alk. paper)
  ISBN-10: 0-8247-5325-9 (alk. paper)
  1. Soil percolation. 2. Soil physics. I. Miyazaki, Tsuyoshi, 1947- II. Series.

S594.M57 2005

Preface to the Second Edition

In revising *Water Flow in Soils*, the original framework of the first edition was kept but several improvements and additions have been made. In Chapter 2, the inverse analyses of infiltration and water flow in deep soils were revised and added. In Chapter 4, the up-to-date subjects relating to fingering flows were added. In Chapter 5, the ponding time on slopes was newly added. In Chapter 6, water flow under temperature gradients in fields and in open soil columns was added. In Chapter 7, recent experimental and theoretical progress in the effects of microbiological factors on water flow in soils were added. In Chapter 8, new topics such as the detection of subsurface water flow by using ^{222}Rn in paddy fields and zero flux plane analysis in upland fields were added. In Chapter 9, the nonsimilar media concept and its extension into unsaturated soils were newly written in detail. Examples and appendixes were added in some chapters to help readers further understand the topics.

I wish to extend my thanks to Russell Dekker, who invited me to publish the first edition and to revise it for the second edition. I also wish to express my gratitude to Katsumi Fujii of the Iwate University for his favorable review, and to Jan W. Hopmans of the University of California, Davis, for his critical review, both addressed to the first edition. These book reviews and even the criticisms encouraged and helped me to revise it. Further, I wish to thank the various publishers and individuals who gave permission for the use of figures and tables of data. Thanks are also due to Hiromi Imoto of The University of Tokyo for his continuous support, and to Moraima Suarez of Marcel Dekker and Patricia Roberson of Taylor & Francis Group for their support in editing the second edition.

The Author

Tsuyoshi Miyazaki, Ph.D., is Professor of Soil Physics and Soil Hydrology at The University of Tokyo, Japan. He is the author or co-author of more than 148 published articles, proceedings, book chapters, and books including *Water Flow in Soils, First Edition* (Marcel Dekker). He received his B.Sc. (1971), M.Sc. (1973), and Ph.D. (1976) from The University of Tokyo. He was a researcher at the Shikoku National Experiment Station (1977–1984), a visiting researcher at the University of California, Davis (1981–1982), a researcher at the National Institute of Agricultural Engineering (1984–1988), and an associate professor at The University of Tokyo (1988–1998).

Dr. Miyazaki is a member of the International Union of Soil Sciences, International Commission of Agricultural Engineering, American Geophysical Union, Soil Science Society of America, and Japanese Society of Irrigation, Drainage, and Reclamation Engineering. He was President of the Japanese Society of Soil Physics (2001–2003) and is on the editorial advisory board of *Soil & Tillage Research.*

Dr. Miyazaki's main research interests are soil physics and environmental soil hydrology.

Contents

1

Soil and Water

I. UNIT SYSTEM

In this book, SI units are used primarily in the definition of physical variables. Prefixes are frequently used with basic SI units. Traditional units are also used, especially when data and figures are quoted from references written using traditional units.

1

The basic SI units are

Length	(m)	meter
Time	(s)	second
Mass	(kg)	kilogram
Amount of substance	(mol)	mole
Temperature	(K)	Kelvin

The units constructed from basic SI units are

Velocity	$(m\,s^{-1})$	
Force	(N)	newton $(m\,kg\,s^{-2})$
Pressure	(Pa)	pascal $(m^{-1}\,kg\,s^{-2})$
Energy	(J)	joule $(m^2\,kg\,s^{-2})$

The appropriate prefixes for use with basic SI units are

Length	(mm)	millimeter
	(cm)	centimeter
	(km)	kilometer
Mass	(g)	gram
	(Mg)	megagram

Traditional units used are

Time	(min)	minute
	(h)	hour
Temperature	(°C)	degree Celsius

The use of SI and traditional units together is not highly desirable but is appropriate given the circumstances noted above.

II.   STRUCTURE OF SOILS

A.   Solid Phase and Soil Pores

Soil is composed of inorganic solid particles of various sizes and irregular shapes. The origins of these particles are weathered rocks, erupted materials, and sediment in ocean, lakes, marshes, and rivers. According to the definition by the

International Society of Soil Science (ISSS), soils are classified by particle size into five classes:

Gravel	larger than 2.0 mm
Coarse sand	between 2.0 and 0.2 mm
Fine sand	between 0.2 and 0.02 mm
Silt	between 0.02 and 0.002 mm
Clay	smaller than 0.002 mm

Typical particle–size distribution curves are given in Figure 1.1 for clay soil, volcanic ash soil, sandy loam, and sand.

Solid particles are composed of primary minerals such as quartz, feldspar, pyroxene, amphibole, and many kinds of mica and secondary minerals such as clay minerals. Clay minerals are predominant in particles smaller than 0.002 mm. Soil texture is defined by the composition of clay, silt, and (fine and coarse) sand. The triangular coordinate as shown in Figure 1.2 is used to classify soils into sand (S), silt loam (SiL), sandy clay (SC), and so on, according to their composition. Soil has various structures, depending on the

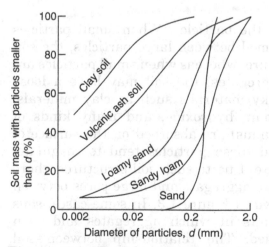

Figure 1.1 Particle–size distribution curves of soils. (From Nakano, M., *Transport Phenomena in Soils,* University of Tokyo Press, Tokyo (1991). With permission.)

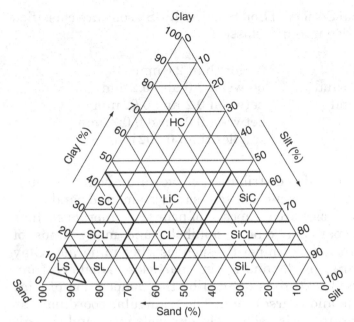

Figure 1.2 Soil texture: S, sand; LS, loamy sand; SL, sandy loam; L, loam; SiL, silt loam; SCL, sandy clay loam; CL, clay loam; SiCL, silty clay loam; SC, sandy clay; LiC, light clay; SiC, silty clay; HC, heavy clay.

spatial arrangement of the particles. When small particles penetrate the pores formed between large particles, the soil may have a dense structure, whereas when small particles are combined and form aggregates, the soil may have a loose structure. Usually, sticky materials such as clay minerals, iron hydroxide, aluminum hydroxide, and many kinds of organic matter (e.g., humus) are absorbed on the surface of large soil particles, and these particles tend to coagulate, resulting in an aggregate. Due to such soil structures, there are small pores within an aggregate and large pores between aggregates, as illustrated in Figure 1.3. In some cases, soils have larger lumped units in which aggregates and even macropores are included. The relationship between soil lump size and surrounding soil type in a field is discussed in Chapter 9.

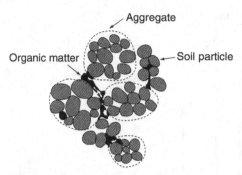

Figure 1.3 Structure of soil aggregate. (After Nakano, M., *Transport Phenomena in Soils,* University of Tokyo Press, Tokyo (1991). With permission.)

The quantities relating to solid phase are defined by the particle density ρ_s ($Mg\,m^{-3}$),

$$\rho_s = \frac{W_s}{V_s} \tag{1.1}$$

by the dry bulk density ρ_d ($Mg\,m^{-3}$),

$$\rho_d = \frac{W_s}{V} \tag{1.2}$$

and by the volumetric solid content σ ($m^3\,m^{-3}$),

$$\sigma = \frac{V_s}{V} \tag{1.3}$$

where V is the total volume of soil, W_s is the mass of soil of volume V, and V_s is the solid volume in volume V, as shown in Figure 1.4. Particle densities, ρ_s, of many soils are almost always 2.65 to 2.70 $Mg\,m^{-3}$. The dry bulk density of sand is generally 1.50 to 1.60 $Mg\,m^{-3}$, and the bulk densities of many other soils are slightly less than this value. The bulk density of Kanto loam, a volcanic ash soil, is about 0.55 $Mg\,m^{-3}$, which is a particularly small value.

The quantities relating to soil pores are defined by the porosity n ($m^3\,m^{-3}$),

$$n = \frac{V_v}{V} \tag{1.4}$$

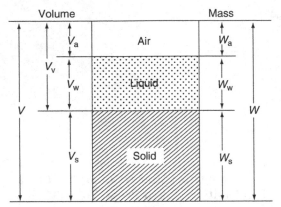

Figure 1.4 Solid, gas, and liquid phases of soil.

and by the void ratio e $(m^3\,m^{-3})$,

$$e = \frac{V_v}{V_s} \tag{1.5}$$

where V_v is the pore volume in volume V. The porosity of soils is generally 0.4 to 0.6 $m^3\,m^{-3}$, while that of Kanto loam is about 0.8 $m^3\,m^{-3}$. The geometry of soil pores is very complicated. This complexity has a significant role on the percolation of materials through the soil.

Specific surface A_s $(m^2\,g^{-1})$, defined by

$$A_s = \frac{a_s}{W_s} \tag{1.6}$$

is another measure of soil, where a_s is the total surface area in volume V. The specific surface of sand is about 0.04 $m^2\,g^{-1}$, about 100 $m^2\,g^{-1}$ for clayey soil, about 300 $m^2\,g^{-1}$ for Kanto loam, about 500 $m^2\,g^{-1}$ for bentonite, and up to 810 $m^2\,g^{-1}$ for pure montmorillonite.

The typical specific surface versus equivalent radius of particles is shown in Figure 1.5 where the index radiuses of particles are given. Note that 1 g of clay has a specific surface almost equal to a tennis court or to the infield of a baseball ground.

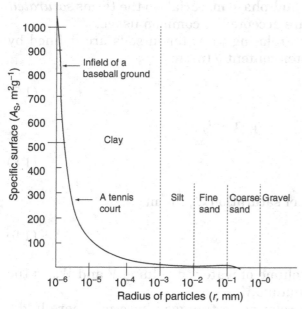

Figure 1.5 Specific surface versus equivalent radius of sphere particles.

Example Derive the specific surface of spherical particles whose radius is $1\,\mu m$ and particle density is $2.65\,Mg\,m^{-3}$.

The mass of one particle M is equal to $4/3\pi r^3 \rho A_s$, and its surface area s is equal to $4\pi r^2$. Therefore, the total surface area of $1\,g$ of this particle A_s is given by

$$A_s = s \times (1/M) = 1.13/r, \quad m^2\,Mg^{-1}$$

Substituting $r = 1\,\mu m = 1 \times 10^{-6}\,m$ in this equation, and changing the unit from Mg to g, the specific surface $A_s = 1.13\,m^2\,g^{-1}$ is obtained.

B. Liquid Phase and Gas Phase in Soil Pores

When soil pores are filled with liquid-phase material, the soil is termed *saturated*; when gas-phase material is contained in the pore spaces, the soil is termed *unsaturated*. In natural fields, however, it is not likely that soil pores will be filled completely with liquid-phase material, so the terms *saturated* and *unsaturated* are accepted in common usage.

The quantities relating to water in soils are defined by the volumetric water content θ ($m^3 m^{-3}$),

$$\theta = \frac{V_w}{V} \tag{1.7}$$

by the water content w ($kg\,kg^{-1}$),

$$w = \frac{W_w}{W_s} \tag{1.8}$$

and by the degree of saturation S_a ($m^3 m^{-3}$),

$$S_a = \frac{V_w}{V_v} \tag{1.9}$$

where V_w is the volume of water in volume V and W_w is the mass of water in volume V.

The quantity relating to the gas phase is generally defined by air-filled porosity a ($m^3 m^{-3}$),

$$a = \frac{V_a}{V} = 1 - \theta - \sigma \tag{1.10}$$

where V_a is the volume of air in volume V.

The property of water in soils differs a little from that of ordinary water. Even when we add pure water to a dry soil, absorbed materials on particle surfaces may be dissolved in the penetrating water, and hence the water is no longer pure but will behave as a solution. A solution that dissolves cations and anions is affected by the negative charge on the surface of clay minerals, resulting in diffuse electrical double layers around the clay minerals due to the attraction of cations and the repulsion of anions. The interaction between water and soil has been discussed in detail by Iwata et al. (1). To define the liquid-phase state more precisely, the potential concept of water in soils will be followed next.

III. RETENTIVITY OF WATER IN SOILS

A. Matric Potential

Water in the vadose zone, the zone between the groundwater level and the land surface, is retained in the soil pores by interaction between the soil particles and water. This interaction reduces the potential energy of water in soils, mainly due to capillarity or adhesion of water molecules to solid surfaces. The decrease in the potential energy of water caused by this interaction is termed the *matric potential* ϕ_m ($\mathrm{J\,kg^{-1}}$), whose values are usually negative in the vadose zone. The *matric head* ψ_m (m) is defined by

$$\psi_m = \frac{\phi_m}{g} \qquad (1.11)$$

where g is the gravitational constant ($9.8\,\mathrm{m\,s^{-2}}$). The *matric potential* is converted into the pressure unit by $\rho_w\phi_m$ (Pa), where ρ_w is the density of water ($\mathrm{kg\,m^{-3}}$). The term *suction*, h (m), given by

$$h = -\psi_m \qquad (1.12)$$

is widely used to designate the absolute value of the matric head ψ_m. It should be noted that matric potential, matric head, matric pressure, and suction can be converted from one to another, as required.

Two simple methods of measuring the matric potential of water in a soil are shown schematically in Figure 1.6: the soil column method in (a) and the tensiometer method in (b). A soil column is placed upright in a water bath as shown in (a) and the water in the soil column is in equilibrium with the water in the bath. A cover is used to prevent the evaporation of water from the top of the column. The vadose zone is above the water level of the bath and the groundwater zone is below it. The matric head of water in the vadose zone at a height z from the water level is given by $-z$. This method is used only when water in the soil column is entirely in equilibrium with water in the bath.

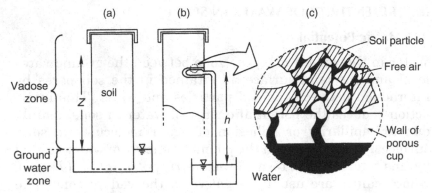

Figure 1.6 Measurement of the matric potential of water in soil by (a) the soil column method, (b) the tensiometer method, and (c) the configuration of water at the soil matrix–porous cup interface.

Another well-known method is the tensiometer method, shown in Figure 1.6(b), where a porous cup filled with water, equilibrated with free water placed below, is inserted into a soil. The state of water at the interface of the soil matrix and the porous cup is illustrated in Figure 1.6(c), where unsaturated water in the soil is in equilibrium with saturated water in the wall of the cup. Generally, an air–water interface is curved and has two main radii of curvature r_1 and r_2. Due to this curvature, the liquid phase has smaller pressure than the ambient air pressure and the difference Δp is given by Laplace's equation,

$$\Delta p = \sigma_w \left(\frac{1}{r_1} - \frac{1}{r_2} \right) \tag{1.13}$$

where σ_w is the surface tension of water.

The matric potential of water in the soil matrix is thus given by measuring the suction h, the vertical distance between the porous cup and the free water placed below. When the water content in the soil decreases, the radius of curvature of the water held in the wall of the cup decreases, and hence the water pressure in the cup decreases, resulting in an increased suction. This method is used at any position in the soil, even in soils under transient states, provided that enough time is allowed for water to be locally equilibrated at the soil matrix–porous cup interface. The sizes of the pores in the wall of the cup limit the feasibility of this method.

It is occasionally emphasized that the air phase in Figure 1.6 must be free and its pressure must be maintained equal to that of atmosphere. When the air pressure differs from atmospheric pressure, we have to define the value of h in Figure 1.6(b) as the tensiometer pressure potential.

B. Soil Moisture Characteristic Curve

The relation between matric potential and volumetric water content in a soil is termed the *soil moisture characteristic curve* because the curve is characteristic of each soil. The curves in Figure 1.7 are the typical soil moisture characteristic curves where matric head ψ_m is used (2). The differences among soil moisture characteristic curves are attributed primarily to the differences in pore size distribution among soils. These curves are sensitive to the changes in bulk densities and disturbances of soil structures. In addition, the curves generally show hysteresis according to the wetting or drying of soils, as described later. It is therefore recommended that

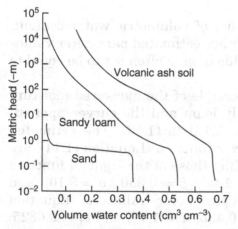

Figure 1.7 Moisture characteristic curves (drying) of some soils. (After Nakano, M., *Transport Phenomena in Soils*, University of Tokyo Press, Tokyo (1991). With permission.)

these conditions be added to each curve accordingly as required.

Soil moisture characteristic curves are approximated by several types of functions. The simplest is given by the power function

$$\psi_m = -a\theta^{-b} \tag{1.14}$$

where a and b are estimated parameter values. The sigmoid function is given by

$$\log(-\psi_m) = a + b\ln\left[\left(\frac{\theta}{\theta_s}\right)^{-c} - 1\right] \tag{1.15}$$

where a, b, and c are estimated parameter values and θ_s the volumetric water content of saturated soil. The function proposed by van Genuchten (3) is given by

$$\Theta = \left[\frac{1}{1 + (\alpha h)^n}\right]^m \tag{1.16}$$

in which the volumetric water content θ is transformed into the dimensionless water content

$$\Theta = \frac{\theta - \theta_r}{\theta_s - \theta_r}$$

where θ_r is the residual value of volumetric water content. In this equation, α, m, and n are estimated parameter values and h is the suction. The value of m is often set to be equal to $1 - 1/n$ for expediency.

Figure 1.8 shows an example of the measured moisture characteristic curve of sandy loam and the curves approximated by using Equations (1.14) to (1.16). The estimated parameter values of the power function (Equation (1.14)) are $a = 0.0385$ and $b = 4.56$, while those of the sigmoid function (Equation (1.15)) are $a = 1.00$, $b = 0.350$, $c = 3.10$, and $\theta_s = 0.420$. The estimated parameter values of Equation (1.16) are $\alpha = 0.0872$, $m = 0.435$, $n = 1.77$, and $\theta_r = 0.0825$. Note that the resultant values of ψ_m with these parameters

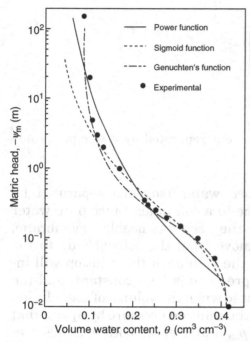

Figure 1.8 Measured and approximated moisture characteristic curves (drying) of sandy loam.

are given in centimeters while they are plotted in meters in Figure 1.8.

As exemplified in Figure 1.8, a power function can cover a wide range of the soil moisture characteristic curve, but its agreement is relatively poor. A sigmoid function is useful in approximating the curve in the vicinity of the inflection point, but its agreement is poor in the case of low water content. Van Genuchten's function is very useful within the range θ_r to θ_s, but the determination of the value of residual value of water content θ_r is not necessarily easy.

Other functions for approximating soil moisture characteristic curves are given in the literature (4).

C. Osmotic Effect on Retentivity

When water in soil contains solute, the matric potential is not sufficient to define the state of water in the soil. Figure 1.9

Semipermeable membrane

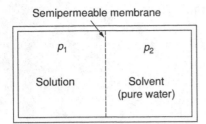

Figure 1.9 Solution and solvent separated by a semipermeable membrane.

shows a solution and pure water (solvent) separated by a semipermeable membrane in a container. Since pure water can permeate through the semipermeable membrane, water molecules tend to move from the solvent side to the solute side. Consequently, the volume of the solution will increase when the solution pressure is kept constant, and the solution pressure will increase when the volume of the solution is kept constant. Denoting the solution pressure by p_1 and that of pure water (solvent) by p_2, and keeping both volumes constant, the osmotic pressure π of the solution is given by

$$\pi = p_1 - p_2 \tag{1.17}$$

It is well known that the chemical potential of water in a solution is smaller than the chemical potential of pure water due to osmotic pressure (1). The higher the osmotic pressure, the smaller the chemical potential value of water in the solution. Thus, the state of water in soils is affected by the osmotic pressure. The decrease in the chemical potential of water caused by osmotic pressure in soil water, termed *osmotic potential* ϕ_o (J kg^{-1}), has negative values. Osmotic head ψ_o (m) is defined by

$$\psi_o = \frac{\phi_o}{g} \tag{1.18}$$

Osmotic pressure π is defined by $\rho_w \phi_o$ (Pa).

The magnitude of the contribution of osmotic potential to the retentivity of water in soils is related to the sizes of soil pores, to the electrical properties of solid surfaces, to the types of ions in the soil solutions, and to their concentrations. For example, when the pore sizes of a soil are larger than water molecules but smaller than hydrated solute molecules, the soil will behave just like the semipermeable membrane in Figure 1.9 and will absorb a large quantity of water under a given pressure, resulting in an increase in water retentivity.

Generally, the pore sizes of sandy soils are considerably larger than those of hydrated solute molecules, while the pore sizes of clay soils are occasionally smaller than those of hydrated solute molecules. The retentivity of water by a soil containing much clay is therefore closely related to the osmotic potential. Swelling of clay soil is brought about by the high retentivity of water due to the osmotic potential as well as by the molecular forces between clay and water and by electrical forces in electrical double layers around clay particles (1,2).

Since natural soils have various pore sizes and soil solutions dissolve several kinds of solutes, the contribution of osmotic potential to the retention of water by soils is determined individually. The contribution of osmotic potential to the flow of water in soils is discussed further in Chapter 6.

D. Hysteresis

Almost all the soil moisture characteristic curves show hysteresis due to the *ink bottle effect*. The simplest model of this effect is given in Figure 1.10, where one thin glass tube and two thin glass tubes with expanded spaces at the halfway point are shown. The inside walls of the glass tubes are assumed to be clean. When an empty tube is placed in a water bath, water will rise up within the tube until the water reaches its equilibrium state (wetting process); while when a tube filled with water is placed in the same water bath, water will be drained until the water reaches its equilibrium state (drying process).

The state of equilibrium achieved through a wetting process and through a drying process may be identical in a straight tube, as shown in Figure 1.10(a), but may be different in tubes with expanded spaces, as shown in Figure 1.10(b),

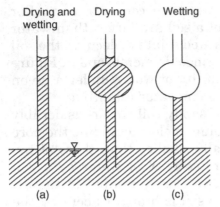

Drying and   Drying   Wetting
wetting

(a)   (b)   (c)

Figure 1.10   Simple model of ink bottle effect in capillary tubes.

where the expanded space is filled with water, and as shown in Figure 1.10(c), where the expanded space is empty.

Since the sizes of soil pores change successively, the existence of the ink bottle effect illustrated above is generally recognized. Figure 1.11 is a typical soil moisture characteristic

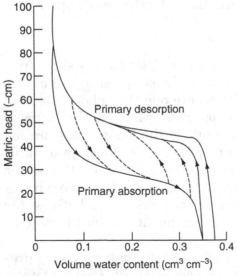

Figure 1.11   Hysteresis of moisture characteristic curve of sand. (After Nakano, M., *Transport Phenomena in Soils,* University of Tokyo Press, Tokyo (1991). With permission.)

curve showing hysteresis. The primary drying curve begins at a saturated condition and extends to a dry condition. The primary wetting curve should be measured by starting from an absolutely dry condition, but practically, it is available by starting from a low matric head where the gradient of the matric head $\partial\psi_m/\partial\theta$ is very large. The primary wetting curve in Figure 1.11 was obtained by starting with $-100\,\mathrm{cm}$ of the matric head. Other dashed lines in Figure 1.11 are scanning curves that turned from drying to wetting, or vice versa. Volumetric water content at a given matric head in the drying curve is larger than in the wetting curve. Physical descriptions of hysteresis are given in detail by Iwata et al. (1).

REFERENCES

1. Iwata, S., Tabuchi, T., and Warkentin, B. P., *Soil–Water Interactions*, 2nd edn, Marcel Dekker, New York (1995).
2. Nakano, M., *Transport Phenomena in Soils,* University of Tokyo Press, Tokyo (1991).
3. van Genuchten, M. Th., A closed-form equation for predicting the hydraulic conductivity of unsaturated soils, *Soil Sci. Soc. Am. J. 44*:892–898 (1980).
4. Kosugi, K., Hopmans, J. W., and Dane, J. H., Parametric Models, in *Methods of soil Analysis, Part 4, Physical Methods*, Dane, J. H., and Topp, G. C. Eds, SSSA, Madison, WI, pp. 739–757 (2002).

2

Physical Laws of Water Flow in Soils

I. DARCY'S LAW

A. Darcy's Equation for Water Flux

Liquid water in a soil tends to move from a location where the potential energy is high to a location where the potential energy is low. The potential energy of water in a soil is termed *total potential* and is composed mainly of matric potential, gravitational potential, and osmotic potential. The former two potentials contribute to the driving forces of water in all soils. The osmotic potential contributes as a driving force, especially in clayey soils. Since almost all natural soils contain a clay component, the osmotic potential has a greater or lesser influence on the flow of water in soils. In moderately moistened natural soils, it is assumed that the contribution of the osmotic potential is negligible and that gravitational and matric potentials predominate in the flow of water. Flow affected by the osmotic potential is discussed in Chapter 6.

The flux of water is given by the Darcy equation

$$q = -K \operatorname{grad} H \tag{2.1}$$

where q is the flux (m s^{-1}), H is the hydraulic head (m), and grad H is the gradient of the hydraulic head. In saturated soils, K is defined as saturated hydraulic conductivity (m s^{-1}); in unsaturated soils, K is the unsaturated hydraulic conductivity (m s^{-1}). By using rectangular coordinate (x,y,z), the components of flux q in the x, y, and z directions are given by

$$q_x = -K_x \frac{\partial H}{\partial x} \tag{2.2}$$

$$q_y = -K_y \frac{\partial H}{\partial y} \tag{2.3}$$

$$q_z = -K_z \frac{\partial H}{\partial z} \tag{2.4}$$

where K_x, K_y, and K_z (m s^{-1}) are either saturated hydraulic conductivity or the unsaturated hydraulic conductivity in the x, y, and z, directions, respectively. The origin of the rectangular coordinate is arbitrary. Usually, the coordinate of the z-axis is set at zero at the land surface, groundwater level, or depth of impermeable layer.

The hydraulic head of water in saturated soil is defined by

$$H = \psi_p + \psi_g \tag{2.5}$$

where ψ_p is the pressure head (m), and the hydraulic head of water in unsaturated soils is defined by

$$H = \psi_m + \psi_g \tag{2.6}$$

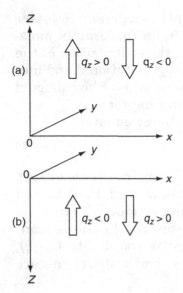

Figure 2.1 (a) Upward positive coordinate system. (b) Downward positive coordinate system.

where ψ_m is the matric head (m) and ψ_g is the gravitational head (m). The components of the flux q in unsaturated soils are then given by

$$q_x = -K_x \frac{\partial \psi_m}{\partial x} \tag{2.7}$$

$$q_y = -K_y \frac{\partial \psi_m}{\partial y} \tag{2.8}$$

$$q_z = -K_z \frac{\partial}{\partial z}(\psi_m + \psi_g) \tag{2.9}$$

Since the physical basis of capillary potential, a major component of matric potential, was given by Buckingham (1), these equations are named Buckingham–Darcy equations. The matric head ψ_m is altered by the pressure head ψ_p in these flux equations for saturated soils.

When the z-axis is defined to be positive vertically upwards (Figure 2.1(a)), Equation (2.9) is rewritten as

$$q_z = -K_z \left(\frac{\partial \psi_m}{\partial z} + 1 \right)$$ (2.10)

when the z-axis is defined to be positive vertically downward (Figure 2.1(b)), the equation is given by

$$q_z = -K_z \left(\frac{\partial \psi_m}{\partial z} - 1 \right)$$ (2.11)

When the suction h (m), whose sign is always positive, is used in these equations, the matric head ψ_m is replaced with $-h$ (m) in these equations. Figure 2.1 denotes, for convenience, the dependence of the sign of the vertical flux q_z on the definition of the z-axis. Occasionally, the z-axis positive upward system is used in the situation when upward flow is predominant and the z-axis positive downward system is used in the situation where downward flow is predominant.

When the z-axis is inclined at an angle ϕ around the y-axis, as shown in Figure 2.2, the hydraulic head H is defined by

$$H = \psi_m - x \sin \phi + z \cos \phi$$ (2.12)

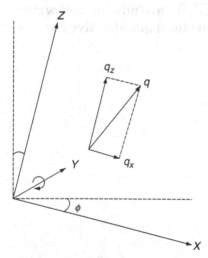

Figure 2.2 Coordinate system rotating the y-axis for ϕ and components of flux q.

and the components q_x and q_z are given by

$$q_x = -K_x\left(\frac{\partial \psi_m}{\partial x} - \sin \phi\right) \qquad (2.13)$$

$$q_z = -K_z\left(\frac{\partial \psi_m}{\partial x} + \cos \phi\right) \qquad (2.14)$$

respectively, while the component q_y remains in the form of Equation (2.8). These equations are used to analyze flow in slopes.

B. Basic Equation of Water Flow in Soils

Liquid water flows through continuous and tortuous pores in soils. Darcy's law is a macroscopic law deduced by the integration of the individual water flow in each pore, which has various shapes microscopically. The basic equation of water flow in soils is constructed by applying Darcy's law to a small cube of $dx\ dy\ dz$ as shown in Figure 2.3, which contains a sufficiently large number of soil particles and therefore demonstrates representative properties of the entire soil sample.

Denoting the change in volumetric water content in this small cube by $d\theta$, the total quantitative change of water in this cube is given by $d\theta\ dx\ dy\ dz$. When there are no source of water and no sink of water in the cube, the total quantitative change

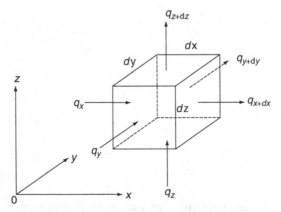

Figure 2.3 Small cube within a soil.

of water is produced by the difference between inflows (q_x, q_y, and q_z) and outflows (q_{x+dx}, q_{y+dy}, and q_{z+dz}), given by

$$q_{x+dx} = q_x + \frac{\partial q_x}{\partial x} dx$$

$$q_{y+dy} = q_y + \frac{\partial q_y}{\partial y} dy \qquad (2.15)$$

$$q_{z+dz} = q_z + \frac{\partial q_z}{\partial z} dz$$

The continuity equation of water during a given time dt is then given by

$$d\theta \, dx \, dy \, dz$$

$$= \left(-\frac{\partial q_x}{\partial x} dx \, dy \, dz - \frac{\partial q_y}{\partial y} dx \, dy \, dz - \frac{\partial q_z}{\partial z} dx \, dy \, dz \right) dt$$

$$(2.16)$$

which results in

$$\frac{\partial \theta}{\partial t} = -\left(\frac{\partial q_x}{\partial x} + \frac{\partial q_y}{\partial y} + \frac{\partial q_z}{\partial z} \right) \qquad (2.17)$$

Applying Darcy's law and substituting Equations (2.2) to (2.4) into Equation (2.17),

$$\frac{\partial \theta}{\partial t} = \frac{\partial}{\partial x}\left(K_x \frac{\partial H}{\partial x} \right) + \frac{\partial}{\partial y}\left(K_y \frac{\partial H}{\partial y} \right) + \frac{\partial}{\partial z}\left(K_z \frac{\partial H}{\partial z} \right) \quad (2.18)$$

is obtained.

In saturated soils, the hydraulic conductivity is assumed to be constant, and hence Equation (2.18) is transformed into

$$\frac{\partial \theta}{\partial t} = K_x \frac{\partial^2 H}{\partial x^2} + K_y \frac{\partial^2 H}{\partial y^2} + K_z \frac{\partial^2 H}{\partial z^2} \qquad (2.19)$$

When the soil matrix is incompressible and isotropic, the left-hand side of Equation (2.19) is zero and $K_x = K_y = K_z$, resulting in

$$\frac{\partial^2 H}{\partial x^2} + \frac{\partial^2 H}{\partial y^2} + \frac{\partial^2 H}{\partial z^2} = 0 \tag{2.20}$$

which is the Laplace equation.

In unsaturated soils, Equation (2.18) is transformed into

$$\frac{\partial \theta}{\partial t} = \frac{\partial}{\partial x}\left(K_x \frac{\partial \psi_m}{\partial x}\right) + \frac{\partial}{\partial y}\left(K_y \frac{\partial \psi_m}{\partial y}\right) + \frac{\partial}{\partial z}\left(K_z \frac{\partial \psi_m}{\partial z}\right) + \frac{\partial K_z}{\partial z} \tag{2.21}$$

which is Richards' equation.

When a relation between volumetric water content θ and matric head ψ_m is given, Equation (2.21) is transformed into equations of either θ or ψ_m. Substituting specific water capacity C, defined by

$$C = \frac{d\theta}{d\psi_m} \tag{2.22}$$

into the left-hand side of Equation (2.21), we obtain the equation of flow with respect to ψ_m as

$$C\frac{\partial \psi_m}{\partial t} = \frac{\partial}{\partial x}\left(K_x \frac{\partial \psi_m}{\partial x}\right) + \frac{\partial}{\partial y}\left(K_y \frac{\partial \psi_m}{\partial y}\right)$$
$$+ \frac{\partial}{\partial z}\left(K_z \frac{\partial \psi_m}{\partial z}\right) + \frac{\partial K_z}{\partial z} \tag{2.23}$$

On the other hand, the substitution of specific water capacity C into the right-hand side of Equation (2.21) gives the equation of flow with respect to θ such that

$$\frac{\partial \theta}{\partial t} = \frac{\partial}{\partial x}\left(D_x \frac{\partial \theta}{\partial x}\right) + \frac{\partial}{\partial y}\left(D_y \frac{\partial \theta}{\partial y}\right) + \frac{\partial}{\partial z}\left(D_z \frac{\partial \theta}{\partial z}\right) + \frac{\partial K_z}{\partial z} \tag{2.24}$$

where D_x, D_y, and D_z are the soil water diffusivities (m^2 s^{-1}), which are defined by

$$D_x = \frac{K_x}{C}, \quad D_y = \frac{K_y}{C}, \quad D_z = \frac{K_z}{C} \tag{2.25}$$

It should, however, be noted that the specific water capacity C, defined by Equation (2.22), is different for each soil, and even in the same soil it is affected by the hysteresis loops of the soil moisture characteristic curve (Figure 1.11).

C. Unsaturated Hydraulic Conductivity

Saturated and unsaturated hydraulic conductivities are both related to the degree of resistance from soil particles when water flows in pores. These resistances are affected by the forms, sizes, branchings, jointings, and tortuosities of pores as well as by the viscosity of water. In addition, unsaturated hydraulic conductivity is affected markedly by the volumetric water content of soil.

There are two types of configurations of water in unsaturated soils: the water films around soil particles, including funicular water at the contacts between particles and the partially saturated water in small pores. These configurations can coexist. When volumetric water content decreases in a soil, both the thickness of water films and the size of domains where partially saturated water exists will decrease. Hence, decease in volumetric water content will bring about a decrease in cross-sectional flow area, an increase in resistance against the flow, and an increase in the real distance of the flow. This is why unsaturated hydraulic conductivity decreases very quickly with a decrease in water content.

Several types of hydraulic conductivities, collected primarily by Nakano (2), are shown as a function of volumetric water content in Figure 2.4. The differences in unsaturated hydraulic conductivities among soils at the same volumetric water content in Figure 2.4 are attributed to the differences in water configurations in individual soils. The unsaturated hydraulic conductivities of soils are thus influenced significantly by disturbance, compaction, or any other cause of change in the configuration of water. The same influence is recognized in saturated hydraulic conductivity.

Empirical formulas of unsaturated hydraulic conductivity have been proposed as functions of θ, h, or ψ_m according to the requirement:

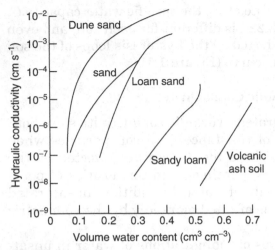

Figure 2.4 Unsaturated hydraulic conductivities of several soils versus volumetric water content. (After Nakano, M., *Transport Phenomena in Soils*, University of Tokyo Press, Tokyo (1991). With permission.)

$$K(\theta) = a\theta^b \tag{2.26}$$

$$K(h) = \frac{a}{b + h^n} \tag{2.27}$$

$$K(\psi_m) = K_s \exp(\alpha\psi_m) \tag{2.28}$$

where a, b, n, and α are parameters determined experimentally.

Many efforts have been made to predict the unsaturated hydraulic conductivities of soils from their moisture characteristic curves, since the curves are related to the structures of soil pores, which dominate the hydraulic properties of soils. Among others, the so-called closed form of unsaturated hydraulic conductivity proposed by van Genuchten (3),

$$K_r(\Theta) = \Theta^{1/2}[1 - (\Theta^{1/m})^m]^2 \tag{2.29}$$

is mostly favored by many researchers, where the relative hydraulic conductivity K_r is defined by the saturated hydraulic conductivity K_s and the unsaturated hydraulic conductivity K by

$$K_r(\Theta) = K/K_s$$

Other parameters in Equation (2.29) are common to those defined in Equation (1.16). To date, however, we do not have a satisfactory method of predicting the unsaturated hydraulic conductivity of a given soil over an entire range of water content. Hence, it is recommended that unsaturated hydraulic conductivities of given soils be measured as precisely as possible.

The effect of hysteresis is negligibly small when unsaturated hydraulic conductivities are given as a function of volumetric water content, whereas the effect of hysteresis is fairly large when the conductivity values are given as a function of matric potential.

The preference for using unsaturated hydraulic conductivities rather than soil water diffusivity in the analysis of water flow in soils is based on evidence that their physical meaning is clear. In addition, water flow in both saturated and unsaturated soils is treated continuously by using unsaturated hydraulic conductivity with saturated hydraulic conductivity. The disadvantage of using unsaturated hydraulic conductivities lies in their difficulty of measurement and in the magnitude of change of their values compared with that of soil water diffusivities.

D. Soil Water Diffusivity

Soil water diffusivity is defined as the ratio of unsaturated hydraulic conductivity to specific water capacity, as given by Equation (2.25). Figure 2.5 shows examples of soil water diffusivities. The advantage of using soil water diffusivities in the analysis of flow lies in the evidence that the range of their values against water content is relatively narrow. This makes numerical treatment of flow easy. The disadvantage of using soil water diffusivities emerges when the flow is analyzed by Equation (2.24), where it is required that both soil water diffusivity and hydraulic conductivity of the given soil be

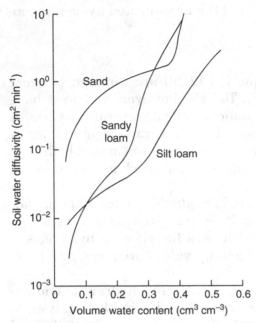

Figure 2.5　Soil water diffusivities of several soils versus volumetric water content. (From Nakano, M., *Transport Phenomena in Soils,* University of Tokyo Press, Tokyo (1991). With permission.)

known. The very small values of specific water capacity C in the range of high water content cause another difficulty when applying soil water diffusivity to soils with a high water content.

II.　GAS-PHASE CONFIGURATION IN UNSATURATED SOILS

A.　Free and Entrapped Air in Soils

There are two types of gas-phase configurations in soils: free air, which is connected with the atmosphere, and entrapped air, which is isolated and surrounded by the liquid phase. The flow of water associated with free and entrapped air is sometimes called flow in an open system and flow in a closed system, respectively (4).

The relation between the soil moisture characteristic curve and gas-phase configuration is illustrated in Figure 2.6, where the soil matrix and water are indicated by shading them together. This type of soil moisture characteristic curve is obtained from both a drying process and a wetting process, each of which is generally different, due to hysteresis (see Chapter 1). Air entry suction h_e (or air entry value ψ_{me}) is defined as the minimum value of suction that allows free air to enter soil pores in the drying process. When the suction is less than h_e, the gas phase can exist only as entrapped air, and when the suction is greater than h_e, free air predominates in the soil pores. When the volume of entrapped air is negligibly small, soil whose suction is less than h_e is considered saturated. Similarly, water entry suction h_w (or water entry value ψ_{mw}) is defined as the maximum value of suction that allows water to enter soil pores when no free air remains in the soil pores during the wetting process. Air entry suction is normally higher than water entry suction. These two types of suction play important roles in preferential water flow (discussed in Chapter 4).

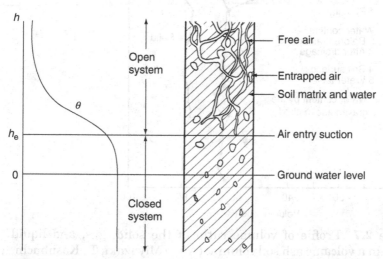

Figure 2.6 Soil moisture characteristic curve and configurations of the gas phase in soils.

Figure 2.7 shows an actual example of volume ratio profiles of solid, gas, and liquid phases in a volcanic ash soil
column 6-cm thick, 10-cm wide, and 108-cm high (5). The
soil column was scanned using a double-gamma-beam
method, with ^{137}Cs and ^{241}Am as sources, to measure the
volumetric water content profiles and bulk density profiles
simultaneously without disturbing the column. The liquid-
phase profile before drainage, denoted as 2 in Figure 2.7,
was obtained by raising the water level. The profile after
drainage, designated 3 in Figure 2.7, was obtained by lowering the water table at the level denoted 5 in the figure. It is
evident in Figure 2.7 that even below groundwater level more
than 10% of the gas phase remained in the soil exclusively as
entrapped air. The air entry suction of this soil is estimated to
be 23 cm.

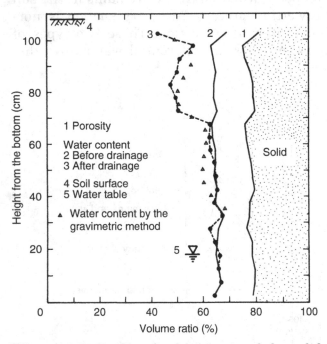

Figure 2.7 Profile of volume ratio of the solid, gas, and liquid
phases in a volcanic ash soil column. (After Miyazaki, T., Kasubuchi,
T., and Hasegawa, S., *J. Soil Sci.* **42**:127–137 (1991). With permission.)

Entrapped air tends to be removed by percolating water when the rate of water flow is large, and tends to be dissolved in water with time, especially when the water is under positive pressure. Hence, air entrapped below the groundwater will diminish over time. On the other hand, the activities of microbes often generate several types of gases in soil pores, which will result in an increase in entrapped air. The change in soil temperature also affects the volume of entrapped air, due to the change in gas solubility in water with temperature. The lower the soil temperature, the lower the volume of entrapped air in soil pores.

Why is the porosity of volcanic ash soil as large as about 80% (see Figure 2.7)? It is quite notable that the porosity of volcanic ash soil is remarkably large compared with other typical soils whose porosities lie in the range 30 to 60%. This peculiarity is due to the predominant clay mineral of volcanic ash soils, Allophane, an amorphous inorganic matter and known as a manifold of a small hollow sphere whose diameter is about 3.5 to 5.0 nm. Since water molecules inside the hollow sphere can be removed away through many holes on the sphere walls, the air-dried volcanic ash soils are fairly light.

B. Forced-Closed System

When channels of air entry into soil pores are interrupted by, say, a tightly closed container, water flows in a forced-closed system even when the suction is greater than the air entry value. Figure 2.8(a) shows horizontal water flow in an open system, and Figure 2.8(b) shows flow in a forced-closed system. The degree of suction on the left-hand side of both columns is maintained at zero, while those on the right-hand side are maintained at h_d. It is assumed that both columns are initially saturated and that h_d exceeds the air entry value of the soil. In the open system, water content will decrease progressively toward the right-hand side of the column by drainage and the unsaturated hydraulic conductivity will be reduced there. In the forced-closed system, even if the suction at the right-hand side of the column exceeds the air entry value, no free air will enter the soil. Assuming that the volume

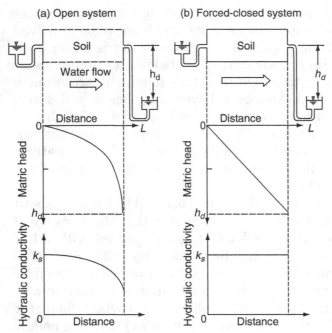

Figure 2.8   Horizontal water flow: (a) in an open system and (b) in a forced-closed system.

of entrapped air is negligible, the hydraulic conductivity in the column will be kept constant throughout the soil. In steady states, therefore, the distribution of matric head in the open system will become a curve that is convex upward while the one in the forced-closed system will become a straight line, as shown in Figure 2.8. In this situation, the flux of water in the forced-closed system will be larger than that in the open system.

C.   Water Flow in Open Systems

By using a steady vertical flow of water in open systems, Srinilta et al. (6) and Nakano and Ichii (7) measured the unsaturated hydraulic conductivities $K(\psi_m)$ of soils as a function of matric head ψ_m. This is a suction method, established by Richards (8). The principle of measurement used by Nakano and Ichii is shown schematically in Figure 2.9,

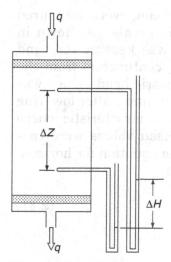

Figure 2.9 Utilization of steady flow in an open system for the measurement of unsaturated hydraulic conductivity.

where a soil sample is installed in a column whose wall is perforated to keep the inside air phase free to the atmosphere. Water flow in soils with such free air phase is called conventionally the flow in an open system. A steady flux q (m s^{-1}) is applied from the upper inlet and matric heads at a distance of Δz along the flow in soil are measured by tensiometers. Applying Buckingham–Darcy equation (2.11) to the various values of q and defining the z-axis as positive vertically downward, the unsaturated hydraulic conductivity $K(\psi_m)$ is given by

$$K(\psi_m) = \frac{q}{-(\Delta\psi_m/\Delta z) + 1} \tag{2.30}$$

where $\Delta\psi_m$ is the difference in matric head at a distance of Δz. The value of the denominator in Equation (2.30) is, by the definition, equal to $\Delta H/\Delta z$, whose value is available by measuring ΔH and Δz in Figure 2.9. The resulting values of $K(\psi_m)$ are given in Figure 2.4.

Miyazaki (9) measured the unsaturated hydraulic conductivities of a sandy loam by using a horizontal transient flow of water in an open system. Matric heads along a perforated soil column, filled with sandy loam, were measured by tensiometers installed at regular intervals, as shown in Figure 2.10. The initial matric head was kept at zero and the outlet was suddenly lowered to z_1 centimeters from the outlet of the column. The change in matric head values was measured at every t_i $(i = 0, 1, 2, \ldots, n)$ minutes after lowering the outlet. By using a soil moisture characteristic curve (which is not indicated), the matric head values were converted to volumetric water content. The equation for horizontal water flow in this column is given by

$$q = -K(\psi_m)\frac{\partial \psi_m}{\partial x} \tag{2.31}$$

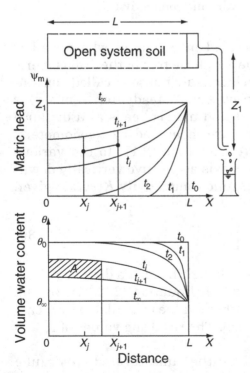

Figure 2.10 Utilization of transient flow in an open system for the measurement of unsaturated hydraulic conductivity.

and the continuity equation of flow is

$$\frac{\partial \theta}{\partial t} = -\frac{\partial q}{\partial x} \tag{2.32}$$

Substituting Equation (2.31) into Equation (2.32) and integrating it from 0 to x, we obtain

$$K(\psi_m) = \left(\frac{\partial}{\partial t} \int_0^x \theta \, dx \right) \left(\frac{\partial \psi_m}{\partial x} \right)^{-1} \tag{2.33}$$

By discretizing time into $t = t_i$ ($i = 0, 1, 2, \dots$) and space into $x = x_j$ ($j = 1, 2, \dots, n$), as shown in Figure 2.10, the value of the right-hand side of Equation (2.33) is estimated approximately. The value of the first parenthetical term on the right-hand side of Equation (2.33) is approximated by

$$\frac{A}{t_{i+1} - t_i} \tag{2.34}$$

where A denotes the change in volume of water per unit section contained within the range $x = 0$ to $x = (x_j + x_{j+1})/2$ between $t = t_i$ and $t = t_{i+1}$. The value of the second parenthetical term on the right-hand side of Equation (2.33) is approximated by

$$\left(\frac{x_{j+1} - x_j}{\psi_{m,j+1} - \psi_{m,j}} \right)_{(t_i + t_i + 1)/2} \tag{2.35}$$

where $(t_i + t_{i+1})/2$ means that the term $(\psi_{m,j+1} - \psi_{m,j})$ is averaged between $t = t_i$ and $t = t_{t+1}$, as denoted in Figure 2.10. The resultant $K(\psi_m)$ obtained by this approximation is also shown in Figure 2.4.

The procedure described here is an instantaneous profile method and is recommended for use with soils that are relatively coarse or loose and thus tend to have less than perfect contact with porous cups during measurement.

D.   Significance of Gas-Phase Configurations
in Fields

Whether the gas phase in a field is free to the atmosphere or entrapped affects the reduction and oxidation of subsoils, their gas-phase composition, denitrification, and the activities of microorganisms. A good example was presented by Tokunaga and Sasaki (10) using forced-closed systems in a test paddy field that had been contaminated by cadmium. It is well known that cadmium is not active and thus is not absorbed by plant roots under reduced conditions. Hence, if water is ponded continuously in a field and subsequently the subsoil is kept under reduced conditions, any influence of the cadmium contamination is blocked. Since the test field soil investigated by Tokunaga and Sasaki (10) was highly permeable it was difficult to maintain ponding, so they crushed and compacted the subsoil to make a forced-closed system. Once such a system was constructed, they could keep the surface soil in reduced conditions and succeeded in prohibiting any influence from the cadmium in the test field.

III.   INFILTRATION

A.   Physical Properties of Infiltration

1.   Infiltration into Dry Soils

Infiltration is defined as the penetration of water into soil pores. If dry sand is spread thinly on a glass plate and water drops are applied to one side of the sand mass by a syringe, water will be absorbed quickly by the sand. Figure 2.11 shows a moment when a distinct wetting front, defined by the interface between a wetted zone (dark part) and a dry zone (light part), exists in glass beads of average diameter 1 mm. The wetting front is moving from the left to the right. Microscopically, however, water does not move smoothly but sometimes accumulates behind the apparently resting wetting front, and then when the equilibrium of the wetting front is destroyed, will suddenly spring into adjacent pores. In Figure 2.11, it is seen that one glass particle is taken in the wetting front in a

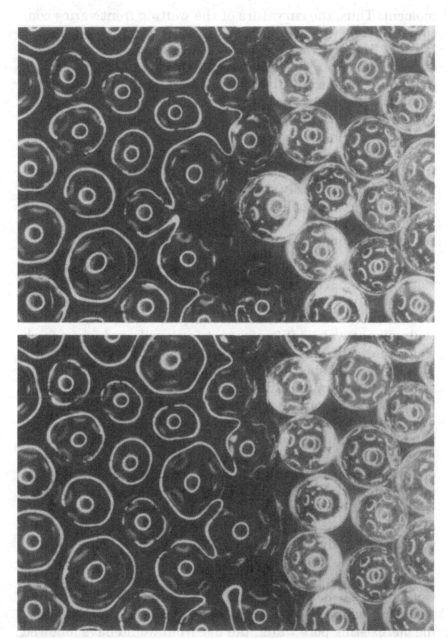

Figure 2.11 Configuration of the wetting front (the interface between the dark zone and the light zone) advancing from left to right in glass beads, taking in the particles one by one.

moment. Thus, the curvature of the wetting front varies continuously and the water film on the wetting front moves laterally rapidly from location to location, due to the difference of pressure. The direction of this film type of flow is parallel to the face of the wetting front, while the direction of the flow behind the wetting front is normal to it.

Although water movement during infiltration is thus very uneven microscopically, the movement appears to be rather smooth when we move our eyes away from the wetting front. Theories of infiltration have been developed primarily for this macroscopically smooth movement of water and partly for the microscopic behavior of the wetting front.

2.   Infiltration into Wet Soils

When new water penetrates wet soils, how does the existing water behave? Will the existing water stay at the surface and at the points of contact of soil particles, or will it be expelled from its positions? This is a matter of great concern both when contaminant water infiltrates wet fresh soils and when fresh water infiltrates wet contaminated soils.

Figure 2.12 shows an experimental device for the demonstration of horizontal infiltration. Sand was packed carefully in a Perspex soil container of 29-cm long, 18-cm wide, and 2-cm deep. Fresh water, whose head was kept at a slightly positive value, was supplied from a reservoir and a dye (a powder of potassium permanganate [$KMnO_4$]) was placed on the surface of the sand bed as a straight line rectangular to the water flow 8 cm from the inlet of water. Figure 2.13 shows the visualized wetting and dye fronts in the sand beds, whose initial water contents were air dried, 0.01 and 0.02 g g^{-1}, respectively. The wetting-front advancement in air-dried sand was identical to that of the dye front (as shown in Figure 2.13(a)), whereas there were differences between the wetting and dye fronts in wet sands (as shown in Figure 2.13(b) and (c)). If infiltrated water moved in the air-filled pores of wet sands without driving out existing pore water, the dye front would have followed the wetting front consistently. On the other hand, if infiltrated water drove out existing pore water, like a piston flow, the dye

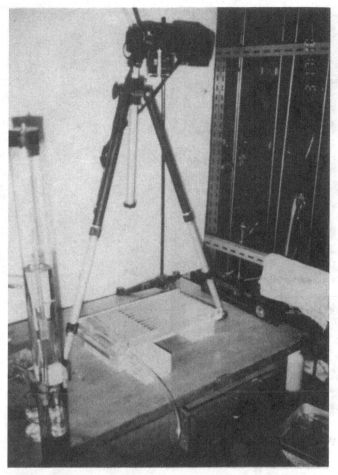

Figure 2.12 Experimental device for the demonstration of horizontal infiltrations.

front would have been behind the wetting front because, according to the assumption, the visible wetting front was not composed of newly infiltrated water but of existing water pushed out by the infiltrated water.

The discrepancies in the wetting and dye fronts in wet sands (Figure 2.13(b) and (c)) show us that the latter assumption made above was true. In other words, water in the vicinity of the wetting front in the wet sand was composed of initially contained water pushed out by newly penetrated

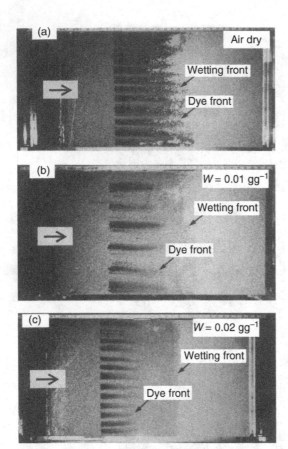

Figure 2.13 Wetting front and dye front in (a) dry sand and (b and c) wet sand.

water. This is fairly important in the understanding of infiltration into wet soils, especially when the infiltrated water is contaminated by chemicals or when fresh water infiltrates wet and polluted soils.

There is a possibility that some of the existing water remains around soil particles during infiltration when soil is highly aggregated or in more clayey soil. van Genuchten (11) termed this remaining water *immobile water*, to distinguish it from mobile water. Determination of the ratio of the water remaining to the water expelled may become a challenging subject in this field.

3. Factors Affecting Infiltration

Infiltration of soils occurs both from ponded water on land surfaces and from the atmosphere in the form of rain or irrigation sprinkling. When the intensity of rain or sprinkling is low, infiltrated water flows down to the soil without ponding on land surfaces.

Figure 2.14 shows examples of simulated water content profiles during infiltration of rains into sandy loam of porosity $0.42 \text{ cm}^3 \text{ cm}^{-3}$ and saturated hydraulic conductivity 2.8×10^{-3} cm s^{-1}. The rain intensities are 2, 5, and 20 mm h^{-1}, respectively. Even when the rain intensity is 20 mm h^{-1} (Figure 2.14(c)), the surface soil is not saturated and the water content gradually approaches $0.26 \text{ cm}^3 \text{ cm}^{-3}$. When the rain intensity is 5 mm h^{-1}, the water content gradually approaches 0.23 cm^3 cm^{-3}. It is supposed that the water content of the surface soil becomes saturated only when the rain intensity exceeds 101 mm h^{-1}, which is equal to the saturated hydraulic conductivity of this sandy loam. When the rain intensity is much greater than this, excess water will pond on the land surface. Thus, infiltrations are classified as either ponded-water infiltration or sprinkling-water infiltration.

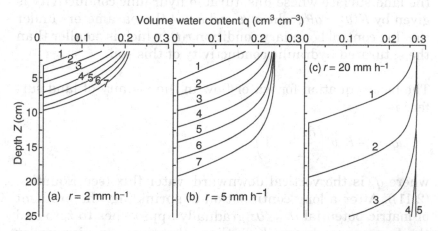

Figure 2.14 Simulated water content profiles during infiltration.

In ponded-water infiltration, the boundary condition of water content at land surface is given by

$$\theta = \theta_s, \qquad z = 0, \ t > 0 \tag{2.36}$$

where θ_s is the saturated volumetric water content. On the other hand, the boundary conditions at the land surface under rain or sprinkling-water infiltration are given by

$$q_0 = r(t), \qquad z = 0, \ t > 0 \tag{2.37}$$

where q_0 is the influx of water and $r(t)$ is the rain or sprinkling intensity. The latter type of boundary condition is termed a flux control boundary condition. Under the flux control boundary condition (2.37), the volumetric water content θ at the land surface is determined by the rain or sprinkling intensity such that the unsaturated hydraulic conductivity $K(\theta)$ is equalized to $r(t)$. Figure 2.14 gives examples where the volumetric water contents of the land surface are controlled by their flux boundary conditions. Initial water content; physical properties of soils relating to swelling, shrinkage, and dispersion of soil particles; heterogeneity of soils; properties of water such as temperature, gas content, and solute concentration; and other environmental conditions all effect infiltration.

Example   Determine the final volumetric water content θ at the land surface whose unsaturated hydraulic conductivity is given by $K(\theta) = a\theta^b$, where a and b are the parameters under the flux control boundary condition $r(t)$, which is smaller than the saturated hydraulic conductivity of this soil.

The basic equation for water flow in the vicinity of land surface is

$$q_z = -K(\theta)\left(\frac{\partial \psi_m}{\partial z} - 1\right)$$

where q_z is the vertical downward water flux (see Equation (2.11)). After a long continuation of sprinkling, the gradient of matric potential $\partial \psi_m / \partial z$ gradually approaches to zero and the flux approaches to $K(\theta)$. Since the given flux is smaller than the saturated hydraulic conductivity, Equation (2.38) is applicable and, from the relation of $r(t) = K(\theta)$, we obtain the final volumetric water content

$$\theta = \left(\frac{r(t)}{a}\right)^{1/b}$$

B. Infiltration Capacity

The infiltration rate is defined by the flux of water across a land surface into the soil. The maximum infiltration rate of a given soil, which is equal to the infiltration rate of ponded water, as shown schematically in Figure 2.15, is termed *infiltration capacity*, designated f_c. The infiltration capacity of dry soil is higher than that of wet soil. At an early stage of infiltration, since the wetting front exists near the land surface, the high matric head gradient at the wetting front gives rise to a high infiltration capacity. The infiltration capacity decreases with time due to advancement of the wetting front into the deep land zone. Eventually, the infiltration capacity approaches asymptotically the final infiltration rate.

It is reasonable to suppose that the final infiltration rate of ponded water is at least equal to the saturated hydraulic conductivity of a given soil because there must be no matric head gradient at the land surface in the final stage of infiltration and hence the rate must be equal to the saturated hydraulic conductivity. This is not the case, however, even under

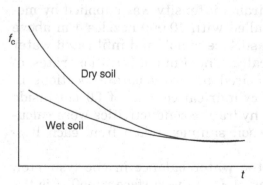

Figure 2.15 Infiltration capacity of dry soil and wet soil.

ideal experimental conditions, and final infiltration rates
have been reported to be one half or two thirds those of the
saturated hydraulic conductivities of uniform soils. One of
the most plausible reasons for this discrepancy is that the
volumes of entrapped air may be different in the apparently
saturated soils and the actually saturated soils. Soil saturation by ponded water proceeds downward from the soil surface, leaving much entrapped air, while the soil saturation in
the measurement of saturated hydraulic conductivity generally proceeds upward by supplying water from the bottom,
minimizing the entrapped air. The difference between the
infiltration capacity and the saturated hydraulic conductivity
of a soil is thus attributed to the difference in the volume of
entrapped air, which obstructs the flow of water in soils.

C. Infiltration Rate in Fields

When the intensity of rain or sprinkling water is less than the
value of the saturated hydraulic conductivity of a given soil,
the infiltration rate of rain or sprinkled water is determined
theoretically by the intensity of rain of sprinkling due to flux
control boundary conditions. Miyazaki (12) investigated the
quantitative relations among the infiltration rates of sprinkling water, the final infiltration rates of ponded water, and the
saturated hydraulic conductivities of the soils by using sloping
lysimeters filled with bare sand, bare volcanic ash soil, sand
covered with grass, and volcanic ash soil covered with grass.
Figure 2.16 shows the equipment used in the experiments.
Artificial rain of the desirable intensity was supplied by mobile rain simulators installed with 10,000 nozzles 4 m above
the lysimeters, and both surface runoff and infiltrated water
were measured periodically. The final infiltration rates of
ponded water were measured at two separate locations in
each lysimeter, using a cylindrical column of 30 cm inside
diameter. The saturated hydraulic conductivities were calculated by averaging four soil samples taken from each lysimeter.

Figure 2.17 shows the water balance in one lysimeter,
where r is the rain intensity, R is the surface runoff, I is the

Figure 2.16 Sloping lysimeters equipped with a mobile rain simulator.

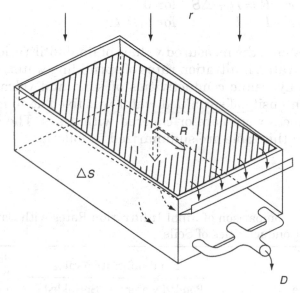

Figure 2.17 Water balance in a sloping lysimeter.

infiltration rate, D is the drainage rate, and ΔS denotes the storage of water in the lysimeter. The water balance changed with time t, as shown in Figure 2.18. When the storage of water in soil was terminated at time t_0, the drainage rate D

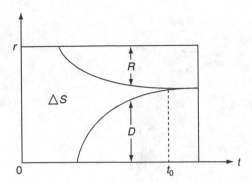

Figure 2.18 Change in water balance.

became equal to the infiltration rate I. The water balance equations in these situations are given by

$$I = \begin{cases} r - R = D + \Delta S & \text{for } 0 < t < t_0 \\ r - R = D & \text{for } t_0 < t \end{cases} \qquad (2.38)$$

Table 2.1 shows the measured values of final infiltration rates of ponded water, infiltration rates of sprinkling water, and the saturated hydraulic conductivities of sand and volcanic ash soil. The intensity of sprinkled water was fixed at 40 mm h^{-1}. Land surfaces were bare or covered with grass. The distinct features in this table are summarized as follows:

Table 2.1 Comparison of Final Infiltration Rates with Saturated Hydraulic Conductivities of Soils

Land surface conditions	Soil	Final infiltration rates		Saturated hydraulic conductivity (mm h^{-1})
		Ponded water (mm h^{-1})	Sprinkled water (mm h^{-1})	
Bare	Sand	43	21	194
	Volcanic ash	858	29	192
Grass	Sand	87	38	124
	Volcanic ash	1175	40	142

1. The final infiltration rates of sprinkled water are much less than both the final infiltration rates of ponded water and their saturated hydraulic conductivities. This feature is attributed to the surface crust formed by the impact of failing drops of water (discussed in more detail in Chapter 5).
2. The final infiltration rates of ponded water into volcanic ash soils are quite large. This is attributed to the macropores (details are given in Chapter 9).
3. Although the rain intensity (40 mm h^{-1}) is lower than the saturated hydraulic conductivities K_s of each soil, the final infiltration rate of sprinkled water was one fourth to one tenth of the K_s value. The values are lower than those reported earlier.
4. The relations among saturated hydraulic conductivities, final infiltration rates of ponded water, and final infiltration rates of sprinkled water are affected by both the physical properties of soils and the land use.
5. The final infiltration rates of ponded water into grass-covered soils were higher than those into bare soils; while the saturated hydraulic conductivities of grass-covered soils were a little lower than those of bare soils. Presumably the scale of macropores and the heterogeneities of the soils are related to these differences.

All these features (Table 2.1) encourage us to understand the hydraulic properties of soils in fields in terms of surface crusts, macropores, and soil heterogeneity (see Chapters 3–5 and 9), and to investigate infiltration more realistically.

D. Mathematical Formulation of Infiltration

1. Empirical Equations

Mathematical formulations of infiltration are classified as empirical methods or physically based methods. Empirical methods try to obtain fitting parameters to approximate the infiltration curve given in Figure 2.15. The Kostiakov equation,

$$i = Bt^{-n} \tag{2.39}$$

the Horton equation,

$$i = i_c + (i_0 - i_c)e^{-kt} \tag{2.40}$$

and the Holtan equation,

$$i = t_c + a(M - I)^n \tag{2.41}$$

are well-known empirical equations in which i is the infiltration rate (identical to the infiltration capacity f_c), i_c is the final infiltration rate, i_0 is the initial infiltration rate, I is the cumulative infiltration, and t is time. The other parameters, denoted by $B, n, k, a,$ and M, are fitting parameters relating to the types of soils and adjacent conditions such as bulk density, initial water content, and nonuniformity of soils.

2. Green–Ampt Equation

The Green–Ampt equation (13) is the first equation of infiltration based on a physical model. In this model, a piston-like moisture profile and a corresponding pressure head profile, designated as solid lines in Figure 2.19, are assumed. The

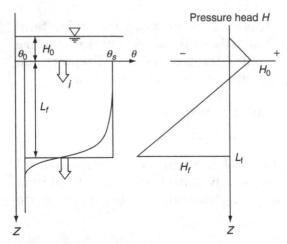

Figure 2.19 Soil moisture profile and pressure head profile of the Green–Ampt model.

smooth curve in Figure 2.19 is the actual moisture profile. In the Green–Ampt model, the infiltration rate is given by

$$i = i_c + \frac{b}{I} \tag{2.42}$$

where b is a physically defined parameter. Alternatively, this equation is written by using saturated hydraulic conductivity K_s and the pressure heads as

$$i = K_s \frac{H_0 - H_f + L_f}{L_f} \tag{2.43}$$

where H_0 is the pressure head at the land surface (which is equal to the depth of ponded water), H_f is the effective pressure head (negative value) at the wetting front, and L_f is the distance from the land surface to the assumed wetting front. The increase in the volumetric water content $\Delta\theta$ in the wetted zone is defined by $\theta_s - \theta_0$, where θ_s is the volumetric water content in the wetted zone and θ_0 is the initial volumetric water content. The value of $\Delta\theta$ is related to the infiltration rate by

$$i = \frac{dI}{dt} = \Delta\theta \frac{dL_f}{dt} \tag{2.44}$$

Substitution of Equation (2.44) into Equation (2.43) yields

$$\frac{K_s}{\Delta\theta} dt = \frac{L_f}{H_0 - H_f + L_f} dL_f \tag{2.45}$$

Integration of Equation (2.45),

$$\int_0^t \frac{K_s}{\Delta\theta} dt = \int_0^{L_f} \frac{L_f}{H_0 - H_f + L_f} dL_f \tag{2.46}$$

yields the solution

$$\frac{K_s}{\Delta\theta} t = L_f - (H_0 - H_f) \ln\left(1 + \frac{L_f}{H_0 - H_f}\right) \tag{2.47}$$

Substitution of Equation (2.47) into the definition of integrated infiltration

$$I = L_f \Delta\theta \tag{2.48}$$

yields the Green–Ampt equation,

$$I = K_s t + A \ln\left(1 + \frac{I}{A}\right) \tag{2.49}$$

where $A = \Delta\theta(H_0 - H_f)$. The reliability of this equation lies in the physical reality of H_f, the effective pressure head at the assumed wetting front. The vagueness of the definition of H_f had lowered the theoretical reliability of the Green–Ampt equation.

Example   Confirm the equality of Equation (2.43) to Equation (2.42).

Since the integrated infiltration I is equal to $\Delta\theta \times L_f$ (see Figure 2.19), L_f is given by

$$L_f = \frac{1}{\Delta\theta}$$

Substituting this relation into Equation (2.43), we obtain

$$i = K_s + \frac{K_s \Delta\theta(H_0 - H_f)}{I}$$

By defining $b = K_s \Delta\theta(H_0 - H_f)$, we find the relation

$$i = K_s + \frac{b}{I}$$

Assuming that $i = i_c = K_s$ for $t \to \infty$, the infiltration rate is written as

$$i = i_c + \frac{b}{I}$$

Several investigations have been conducted to verify the physical meaning of H_f. Bouwer (14) proposed the critical pressure head concept, with which he estimated the value of H_f, and Mein and Larson (15) developed the critical pressure head concept more realistically. The explanation given by

Neuman (16) of the physical meaning of H_f may be typical. He applied a one-dimensional Darcy's equation by using pressure head H_p as

$$q = -K\left(\frac{\partial H_p}{\partial z} - 1\right) \tag{2.50}$$

where K is the unsaturated hydraulic conductivity. The value of pressure head H_p is equal to the matric head ψ_m when the value is negative. Integration of Equation (2.50) from the land surface to the wetting front is given by

$$q \int_0^{L_f} dz = -\int_{H_0}^{H_i} K \, dH_p + \int_0^{L_f} K \, dz \tag{2.51}$$

where H_0 is the pressure head at the land surface and H_i is the initial pressure head of the soil. When a piston-like moisture profile is assumed during infiltration, the flux q is equal to the infiltration rate i and the unsaturated hydraulic conductivity K is equal to the saturated hydraulic conductivity K_s. The implementation of integration is thus given by

$$i = q = \frac{1}{L_f}\left(-\int_{H_0}^{H_i} K \, dH_p + K_s L_f\right) \tag{2.52}$$

The integration of K from H_0 to H_i is divided into two parts:

$$\int_{H_0}^{H_i} K \, dH_p = \int_{H_0}^{0} K \, dH_p + \int_{0}^{H_i} K \, dH_p \tag{2.53}$$

where the value of H_0 is positive and the value of H_i is negative. The first integration on the right-hand side of Equation (2.53) is equal to $-K_s H_0$ because the value of K is usually equal to K_s under a positive pressure head. Thus, the infiltration rate is given by

$$i = K_s \frac{H_0 - \int_0^{H_i} (K/K_s) \, dH_p + L_f}{L_f} \tag{2.54}$$

Comparing each term of Equation (2.54) with the Green–Ampt equation (2.43), we obtain

$$H_f = \int_0^{H_i} \frac{K}{K_s} dH_p \tag{2.55}$$

which is a physically based definition of the effective pressure head H_f at the wetting front.

The Green–Ampt equation (first published in 1911) was deduced from an oversimplified moisture profile model, a piston-like profile. Nevertheless, this equation is still used because its ability to predict the infiltration rate is no poorer than that of newer equation (17) and because the simplicity of the piston-like moisture profile is favored by engineers and some scientists. Iwata et al. (18) provide further detailed applications of the Green–Ampt model.

3. Philip's Method

A one-dimensional flow equation for horizontal flow in an isotropic soil is given by

$$\frac{\partial \theta}{\partial t} = \frac{\partial}{\partial x_1} \left(D \frac{\partial \theta}{\partial x_1} \right) \tag{2.56}$$

for a vertical-down flow by

$$\frac{\partial \theta}{\partial t} = \frac{\partial}{\partial x_2} \left(D \frac{\partial \theta}{\partial x_2} \right) - \frac{\partial K}{\partial x_2} \tag{2.57}$$

and for a vertical-up flow by

$$\frac{\partial \theta}{\partial t} = \frac{\partial}{\partial x_3} \left(D \frac{\partial \theta}{\partial x_3} \right) + \frac{\partial K}{\partial x_3} \tag{2.58}$$

where θ is the volumetric water content, t is time, x_1 is the horizontal distance, x_2 is the vertical distance positive downward, x_3 is the vertical distance positive upward, K is the unsaturated hydraulic conductivity, and D is the soil water diffusivity. Approximate solutions of these equations for semi-infinite soil under the initial and boundary conditions

$$\theta = \begin{cases} \theta_i & t = 0, \quad 0 \le x_n \\ \theta_i & t > 0, \quad x_n = +\infty \\ \theta_0 & t > 0, \quad x_n = 0 \end{cases} \tag{2.59}$$

for $n = 1$, 2, or 3 were given analytically by Philip (19,20), where θ_i is the initial volumetric water content and θ_0 is the volumetric water content at the soil surface. To date, modifications of Philip's solutions to a variety of conditions have been developed primarily by Philip himself. The concept involved in Philip's solution is introduced here in a specific form.

Miyazaki et al. (21) measured moisture profiles during horizontal, vertical-up, and vertical-down infiltrations. Figure 2.20 shows the measured profiles 60, 120, and 240 min, respectively, after the start of infiltration into air-dried Hanford sandy loam. The boundary conditions of the matric head at the inlet of each soil column were maintained at -5 cm. The distances $x_1(\theta,t)$ of horizontal infiltration, $x_2(\theta,t)$ of vertical-down infiltration, and $x_3(\theta,t)$ of vertical-up infiltration from a cross section of the inlet of water were measured with time using gamma-beam equipment.

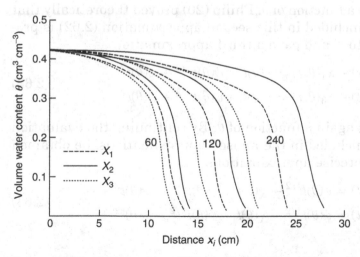

Figure 2.20 Moisture profiles during horizontal, vertical-up, and vertical-down infiltrations. (Data from Miyazaki, T., Nielsen, D.R., and MacIntyre, J.L., *Hilgardia* 52(6):1–24 (1984). With permission.)

It is confirmed by Figure 2.20 that in case of horizontal infiltration, the distance $x_1(\theta,t)$ where a designated water content locates in proportional to the square root of time

$$x_1(\theta, t) = \phi(\theta)t^{1/2} \tag{2.60}$$

where $\phi(\theta)$ is a function of θ only. Discrepancies between $x_1(\theta, t)$ and $x_2(\theta,t)$ and those between $x_1(\theta,t)$ and $x_3(\theta,t)$ are attributed to the effect of gravity. Investigating Figure 2.20, it is likely that these discrepancies increase with time. Hence, as a first approximation, it is reasonable to assume that

$$x_2(\theta, t) - x_1(\theta, t) = \chi(\theta)t$$
$$x_1(\theta, t) - x_3(\theta, t) = \chi(\theta)t \tag{2.61}$$

where χ is a function of θ. Philip (19) proved theoretically that the error included in this first approximation (2.61) is proportional to $t^{3/2}$. Instead of Equation (2.61), he gave a second approximation by

$$x_2(\theta, t) - x_1(\theta, t) = \chi(\theta)t + \psi(\theta)t^{3/2}$$
$$x_1(\theta, t) - x_3(\theta, t) = \chi(\theta)t - \psi(\theta)t^{3/2} \tag{2.62}$$

where ψ is a function of θ. Philip (20) proved theoretically that the error included in this second approximation (2.62) is proportional to t^2 and gave a third approximation as

$$x_2(\theta, t) - x_1(\theta, t) = \chi(\theta)t + \psi(\theta)t^{3/2} + \omega(\theta)t^2$$
$$x_1(\theta, t) - x_3(\theta, t) = \chi(\theta)t - \psi(\theta)t^{3/2} + \omega(\theta)t^2 \tag{2.63}$$

where ω is again a function of θ. By continuing the evaluation of errors included in the revised approximations, he obtained the more precise approximations

$$x_2(\theta, t) = \phi(\theta)t^{1/2} + \chi(\theta)t + \psi(\theta)t^{3/2} + \omega(\theta)t^2 + \cdots$$
$$x_3(\theta, t) = \phi(\theta)t^{1/2} - \chi(\theta)t + \psi(\theta)t^{3/2} - \omega(\theta)t^2 + \cdots \tag{2.64}$$

These series, known as Philip's solution, in fact, give the approximate solutions for nonlinear partial differential equations (2.57) and (2.58), respectively.

Integration of $x_2(\theta,t)$ gives the cumulative infiltration I of vertical-down infiltration,

$$I(t) = \int_{\theta_i}^{\theta_0} x_2(\theta, t) d\theta \tag{2.65}$$

and differentiation of Equation (2.65) by t gives the infiltration rate i,

$$i = \frac{dI}{dt} \tag{2.66}$$

Substituting the approximation for vertical-down infiltration, Equation (2.64), into Equation (2.65) and differentiating it with time t, the infiltration rate is transformed into

$$i = \frac{t^{-1/2}}{2} S + i_c \tag{2.67}$$

where S is the sorptivity,

$$S = \int_{\theta_i}^{\theta_0} (\theta) \, d\theta \tag{2.68}$$

and i_c is

$$i_c = \int_{\theta_i}^{\theta_0} \chi(\theta) \, d\theta + \frac{3t^{1/2}}{2} \int_{\theta_i}^{\theta_0} \psi(\theta) \, d\theta + \cdots \tag{2.69}$$

Equation (2.67) is Philip's infiltration rate. Examples of the values of ϕ, χ, and ψ of Hanford sandy loam, calculated from Figure 2.20, are shown in Figure 2.21.

By integrating $x_1(\theta,t)$ and $x_3(\theta,t)$ from θ_i to θ_0, respectively, in the same manner as in Equation (2.65), cumulative infiltrations and infiltration rates of horizontal and vertical-up infiltrations are also calculated. Although the applicability of Philip's method is restricted to particular initial and boundary conditions and to uniform soils, it provides useful analytical solutions for infiltration.

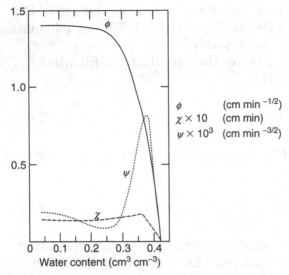

Figure 2.21 ϕ, χ, and ψ values of Hanford sandy loam, calculated from Figure 2.20. (After Miyazaki, T., Nielsen, D.R., and MacIntyre, J.L., *Hilgardia* 52(6):1–24 (1984). With permission.)

4. Inverse Analysis of Infiltration by Using Philip's Solution

Generally, the inverse analysis provides the estimation of unknown input parameters from measured experimental values. Philip's solution of infiltration is capable of estimating the unknown parameters such as soil water diffusivity $D(\theta)$ and unsaturated hydraulic conductivity $K(\theta)$ from measured values of $x_1(\theta,t)$, $x_2(\theta,t)$, and $x_3(\theta,t)$.

Philip (19,20) derived the relation between coefficients $\phi(\theta)$, $\chi(\theta)$ and parameters $D(\theta), K(\theta)$ as

$$\int_{\theta_1}^{\theta} \phi \, d\theta = -2D\frac{d\theta}{d\phi} \tag{2.70}$$

and

$$\int_{\theta_i}^{\theta} \chi \, d\theta = K - K_i - \frac{1}{2}\frac{d\chi}{d\phi}\int_{\theta_i}^{\theta} \phi \, d\theta \tag{2.71}$$

where K_i is the unsaturated hydraulic conductivity for the initial water content θ_i. When the soil is initially air dried, the value K_i is negligibly small. If we determine the values of ϕ and χ as functions of θ from measured values of x_1, x_2, and x_3, the unknown parameters D and K may be estimated by these two equations.

Miyazaki et al. (21) first determined the value of ϕ from Equation (2.60) directly. The value χ was approximated by using Equation (2.64) as

$$\chi = \frac{x_2 - x_1}{2t} + \varepsilon \tag{2.72}$$

where ε is the error caused by the truncation of terms below the fourth. The value of ψ, which is not used in this inverse analysis, was also approximated by

$$\psi = \frac{x_2 + x_3 - 2x_1}{2t^{3/2}} + \varepsilon' \tag{2.73}$$

where ε' is again the error caused by the truncation of terms below the fourth. Figure 2.21 was thus obtained.

Figure 2.22 is the resultant unsaturated hydraulic conductivities of (a) Hanford sandy loam, (b) Yolo light clay, (c) Monona silt loam, and (d) Ida silt loam. The independently measured unsaturated hydraulic conductivities are also plotted in these figures showing good agreements with predictions by this inverse analysis.

IV. STEADY WATER FLOW IN SOILS

A. Steady Flows in Fields

The *steady state of water flow* is defined as the state where water is moving continuously without storage or consumption in the soil. Generally, saturated flows in groundwater and in vadose zones whose suction is less than the air entry value are regarded as steady flow provided that the boundary conditions for flow do not fluctuate practically. Continuous vertical-down water flows under ponded water at land surfaces, and lateral flows of groundwater are typical steady flows in the field.

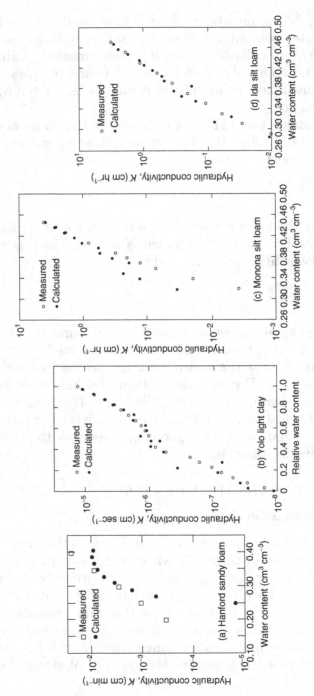

Figure 2.22 Predicted and measured unsaturated hydraulic conductivities. (After Miyazaki, T., Nielsen, D.R., and MacIntyre, J.L., *Hilgardia* 52(6):1–24 (1984). With permission.)

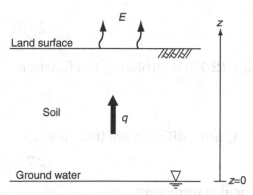

Figure 2.23 Steady upward flow of water during evaporation from land surface.

A completely steady flow in unsaturated soil is generated only in the laboratory, where the boundary conditions of flow are flexible. Flows of water in unsaturated soils are almost always in unsteady states in natural fields due to the changes in boundary conditions for the flows, the changes in water storage in soil pores, and the consumption of soil water by plant roots. The land surface, which is always subject to meteorological fluctuation, is a typical changeable boundary condition for flow in unsaturated soil.

B. Steady Upward Flow

Gardner (22) showed some steady-state solutions for the unsaturated water flow equation with applications to evaporation from a groundwater table, as illustrated in Figure 2.23, where a steady upward flow of water, designated q, occurs from the water table to the land surface. The evaporation rate E at the land surface is assumed to be constant. Darcy's law for this flow is given by

$$q = -K\left(\frac{d\psi_m}{dz} + 1\right) \tag{2.74}$$

where q is the steady upward flux of water whose value is equal to E and the matric head ψ_m is a function of the height z. Integration of z from 0 to ψ_m yields

$$z = -\int_0^{\psi_m} \frac{d\psi_m}{1 + q/K} \qquad (2.75)$$

Gardner (22) solved Equation (2.75) by utilizing the function

$$K = \frac{a}{(-\psi_m)^n + b} \qquad (2.76)$$

in which $n = 1, \frac{3}{2}, 2, 3,$ and 4, and utilizing another function,

$$K = a \exp(c\psi_m) \qquad (2.77)$$

where a, b, and c are empirical parameters.

As an example, let us chose $n = 2$ in Equation (2.76). The denominator in Equation (2.75) is then given by

$$1 + \frac{q}{K} = \beta + \alpha\psi_m^2 \qquad (2.78)$$

where

$$\alpha = \frac{q}{a} \quad \text{and} \quad \beta = 1 + \alpha b$$

and the solution is given by

$$z = (\alpha\beta)^{-1/2} \arctan\left[-\left(\frac{\alpha}{\beta}\right)^{1/2}\psi_m\right] \qquad (2.79)$$

The value of n in Equation (2.76) is different for each soil. Roughly speaking, n is about 2 for clayey soils, about 3 for loamy soils, and increases with soil particle size.

When we chose $n = 3$, the integration of Equation (2.75) yields

$$z = \frac{1}{\alpha}\left\{\frac{1}{6\gamma^2}\ln\left(\frac{(\gamma - \psi_m)^2}{(\gamma^2 + \gamma\psi_m + \psi^2)}\right) + \frac{1}{\gamma^2\sqrt{3}}\arctan\frac{(-2\psi_m - \gamma)}{\gamma\sqrt{3}}\right\} \qquad (2.80)$$

where a displacement $\gamma^3 = \beta/\alpha$ was used. When we chose $n = 4$, the integration yields

$$z = \frac{1}{2\sqrt{2}\alpha\tau^{3/4}} \left[\frac{1}{2} \ln \left(\frac{\psi_m^2 - \sqrt{2}\psi_m\tau^{1/4} + \tau^{1/4}}{\psi_m^2 + \sqrt{2}\psi_m\tau^{1/4} + \tau^{1/2}} \right) \right.$$

$$\left. + \arctan \frac{-2\sqrt{2}\psi_m\tau^{1/4}}{\tau^{1/2} - \psi_m^2} \right] \tag{2.81}$$

where a displacement $\tau = \beta/\alpha$ was used.

Example   Illustrate the matric head distribution under the steady evaporation rate of 1 mm day^{-1} from a bare loam soil surface when the groundwater level is -17 cm and the hydraulic parameters are $n = 2$, $a = 1.8 \times 10^{-4}$, and $b = 1.8$ in Equation (2.76).

Since the upward steady flux q, equal to the steady evaporation rate, is 1.16×10^{-6} cm s^{-1} and since the values of parameters α and β are easily obtained, Equation (2.79) results in

$$\psi_m = -12.5 \tan \frac{z}{12.4}$$

The matric head distribution is given in Figure 2.24.

Solutions to the steady flow equation (2.75) in a clay soil and in a loam soil calculated by Hasegawa (23) are shown in Figure 2.25, where the unsaturated hydraulic conductivity K (cm s^{-1}) of clay soil was approximated by

$$K = \frac{1.8 \times 10^{-4}}{(-\psi_m)^2 + 1.8} \tag{2.82}$$

and that of loam soil was approximated by

$$K = \frac{7}{(-\psi_m)^3 + 2300} \tag{2.83}$$

Figure 2.25 shows solutions for loam soil when the evaporation rates are 3, 6, and 10 mm day^{-1}, respectively, and for clay soil when evaporation rates are 1, 3, and 6 mm day^{-1}, respectively. For example, to generate a steady evaporation rate of 3 mm day^{-1}, the depth of the groundwater table must

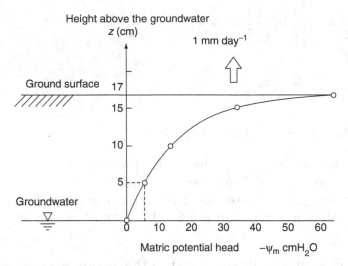

Figure 2.24 Matric head distribution under the steady evaporation of $1 \, \text{mm} \, \text{day}^{-1}$.

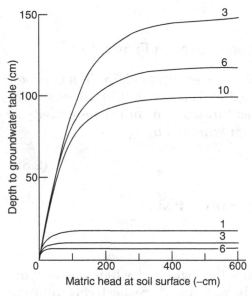

Figure 2.25 Solutions of Equation (2.75) for steady upward flow in loam (upper three curves) and in clay (lower three curves). (After Hasegawa, S., *Soil Phys. Cond. Plant Growth Jpn.* *53*:13–19 (1986). With permission.)

be less than about 150 cm in loam soil and less than about 15 cm in clay soil. Within these ranges, the matric heads at any depth from the water table to the soil surfaces are given by solving Equation (2.75) under steady-state conditions. If the depth of the groundwater table exceeds these limitations, transient evaporation will be generated instead of steady evaporation.

Integration of Equation (2.74) can be carried out not only from $z = 0$ to z but also from an arbitrary z value to another z value. Willis (24) analyzed steady evaporations in layered soils using this character of integration.

C. Steady Downward Flow

Steady downward flow of water is also described by Darcy's law, Equation (2.74), provided that q is negative in the space where z is positive upward. Raats (25) discussed the general features of the integration (2.75) for steady downward flows. Srinilta et al. (6), on the other hand, analyzed steady downward flow in a two-layer soil by integrating Equation (2.74) in each layer. Warrick and Yeh (26) and Warrick (27) reviewed all the numerical approximations concerned with the integration method.

Figure 2.26 is the solution of steady downward flow in a uniform loamy sand (a) and in a layered loamy sand sandwiching a gravel layer (b). The solutions agreed well with the corresponding experimental values plotted in the figure. The matric head profiles can be transformed into water content profiles by using moisture characteristic curves, see Figure 1.8, resulting in Figure 2.27(a) and (b).

In a deep uniform soil or in a very thick subsoil in a layered soil, the upper part of the matric head profiles in the subsoil are vertical straight lines under steady downward flow. The greater the downward flux, the longer the zone of vertical straight line of matric head profile, as shown schematically in Figure 2.28. This is the reason that we often recognize vertically constant matric head profiles in the subsoil of paddies, where ponded water is percolating downward through a relatively impermeable topsoil layer. Readers are

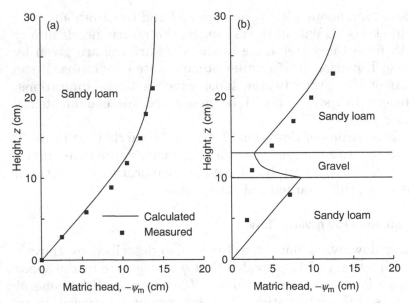

Figure 2.26 Solutions of Equation (2.75) for steady downward flow in (a) a uniform sandy loam and (b) a layered sand loam sandwiching a gravel layer.

referred to the book published by Iwata et al. (18) for further details on the downward percolation of water in paddies.

V. TRANSIENT WATER FLOW IN SOILS

A. Evaporation

1. Fundamental Pattern of Evaporation

Evaporation from soil induces upward soil water movement directed toward the evaporating surface. The maximum evaporation rate (i.e., potential evaporation rate) is dominated by such conditions as temperature, relative humidity, and wind velocity, while the maximum upward flux of soil water is dominated by the hydraulic properties of soil, such as unsaturated hydraulic conductivity and the water potential gradient in the soil. The actual evaporation rate is determined by both the external evaporativity and the hydraulic properties of the soil.

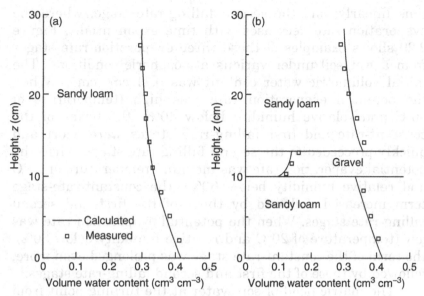

Figure 2.27 (a and b) Soil moisture profiles converted from Figure 2.26.

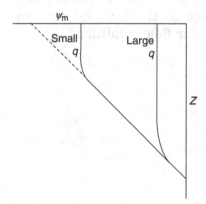

Figure 2.28 Matric head profiles in deep soil with a small downward flux and a large downward flux.

The fundamental evaporation-rate pattern is composed of the following three stages: the constant-rate stage, where the evaporation rate does not change with time; the first falling-rate stage, where the evaporation rate decreases with time linearly; and the second falling-rate stage, where the evaporation rate decreases with time exponentially. Figure 2.29 shows examples of these three evaporation-rate stages from a wet soil under various atmospheric conditions. The initial volumetric water content was 0.41 cm^3 cm^{-3}. When the potential evaporation rate was high (temperature of 30°C and relative humidity below 40%), the terms of the constant-rate and first falling-rate stages were short and quickly proceeded to the second falling-rate stage. When the potential evaporation rate was medium (temperature of 20°C and relative humidity below 60%), the constant-rate-stage term increased, followed by those of the first and second falling-rate stages. When the potential evaporation rate was low (temperature of 20°C and relative humidity below 80%), the term of the constant-rate stage was prolonged even more, followed by those of the first and second falling-rate stages.

The matric head of soil water at the turning point from the constant-rate stage to the first falling-rate stage and that from the first falling-rate stage to the second falling-rate stage are determined by the potential evaporation rate and hydraulic properties of the soil. The matric head at the turning point from the constant-rate stage to the first falling-rate stage is generally about −10 m under field conditions.

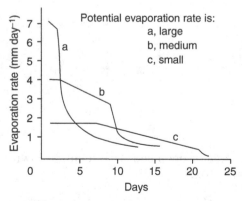

Figure 2.29 Actual evaporation rates under different potential evaporation rates. (After Nakano, M., Miyazaki, T., and Maeda, S., *Soil Phys. Cond. Plant Growth Jpn. 58*:30–39 (1988). With permission.)

2. Transient Water Flow in the Constant-Rate Stage

Roughly speaking, when the evaporation rate E is much less than the unsaturated hydraulic conductivity of surface soil K_0, constant-rate evaporation is likely to occur. Nakano (29) analyzed the transient upward flow of water in the constant-rate stage using the condition

$$E \ll K_0 \tag{2.84}$$

The one-dimensional vertical flow equation is given by the continuity condition of flow

$$\frac{\partial \theta}{\partial t} = -\frac{\partial q_z}{\partial z} \tag{2.85}$$

where q_z is the vertical flux given by Equation (2.4). Defining a new variable ζ by

$$\zeta = \frac{\partial q_z}{\partial \theta} = \frac{\partial}{\partial \theta}\left[-K\left(\frac{\partial H}{\partial z}\right)\right] \tag{2.86}$$

where H is the hydraulic head $(= \psi_m + z)$, Equation (2.85) is transformed into the well-known first-order partial differential equation

$$\frac{\partial \theta}{\partial t} = \zeta\frac{\partial \theta}{\partial z} = 0 \tag{2.87}$$

This is a kinematic wave equation with the propagation velocity ζ. Solution of the equation (see the appendix to this chapter)

$$\frac{dz}{dt} = \zeta \tag{2.88}$$

yields

$$(t + A)\zeta - z = B \tag{2.89}$$

where A and B are integral constants. Substituting Equation (2.86) into Equation (2.89) and integrating, we obtain

$$-\frac{\partial H}{\partial z} = \frac{(z+B)(\theta+C)}{K(t+A)} \tag{2.90}$$

where C is a constant of integration.

Nakano (29) solved Equation (2.90) under the initial and boundary conditions

$$-\frac{\partial H}{\partial z} = \begin{cases} 0, & t=0,\ z\le 0 \\ \frac{E}{K_0}, & t>0,\ z=0 \\ & t>0,\ z=-L \end{cases} \tag{2.91}$$

where z is zero at the soil surface and L is the depth of soil. Determining the constants of integration in Equation (2.90) from the conditions (2.91), the solution is rewritten as

$$-\frac{\partial \psi_m}{\partial z} = 1 + \frac{E}{K_0}\frac{L+z}{L}\frac{\theta(z,t)-\theta(z,0)}{\theta(0,t)-\theta(0,0)} \tag{2.92}$$

By using assumption (2.84), the second term of Equation (2.92) is eliminated, resulting in

$$\frac{\partial \psi_m}{\partial z} = 1 \tag{2.93}$$

Nakano (29) estimated that this assumption is applicable when the condition

$$E_p < K_0/5 \tag{2.94}$$

is satisfied, where E_p is the potential evaporation rate.

Figure 2.30 shows the change of matric head profile during constant-rate evaporation obtained by solving Equation (2.93). Note that all the profiles have the same gradient in this situation. The change in moisture profile is also calculated by converting the solution of Equation (2.93) into soil water content. Nakano (29) approximated a soil moisture characteristic curve by

$$\psi_m = -\exp(a-b\theta) \tag{2.95}$$

where a and b are the empirical constants of soil. Figure 2.31 shows the measured moisture profiles and the theoretically predicted profiles. The agreement among these profiles is

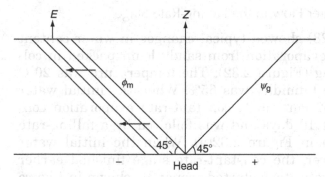

Figure 2.30 Change of matric head profile during constant-rate evaporation.

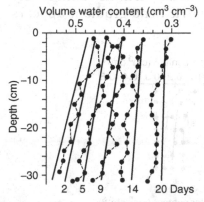

Figure 2.31 Measured and predicted moisture profiles under constant-rate evaporation. (After Nakano, M., *Soil Sci.* 124(2):67–72 (1977). With permission.)

excellent. When condition (2.94) is not satisfied, the theoretically predicted profiles of matric heads and water content will gradually deviate from the experimental values. Upward flow of water during constant-rate evaporation is thus characterized by relatively large unsaturated hydraulic conductivities of soils and by linear profiles of matric heads in soils.

3. Transient Water Flow in the Falling-Rate Stage

Nakano et al. (28) showed typical changes in water content profiles during evaporation from sandy loam packed in columns 14-cm long (Figure 2.32). The temperature was 20°C and the relative humidity was 65%. When the initial water content was 0.52 cm^3 cm^{-3}, constant-rate evaporation continued for about 10 days and was followed by a falling-rate stage, as shown in Figure 2.32(a). When the initial water content was lower, the constant-rate stage finished earlier and the falling-rate stage started sooner, as shown in Figure 2.32(b) to (d).

During constant-rate evaporation, unsaturated hydraulic conductivity of the soil decreases due to the lower water content. When the unsaturated hydraulic conductivity is less than about $E_p/5$, a large matric head gradient appears

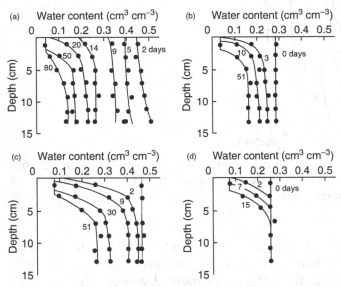

Figure 2.32 Changes in moisture profiles during falling-rate evaporation. The initial volumetric water contents are 0.52 cm^3 cm^{-3} (a), 0.29 cm^3 cm^{-3} (b), 0.23 cm^3 cm^{-3} (c), and 0.13 cm^3 cm^{-3} (d), respectively. (After Nakano, M., Miyazaki, T., and Maeda, S., *Soil Phys. Cond. Plant Growth Jpn.* 58:30–39 (1988). With permission.)

in the vicinity of the soil surface in order to supply water within the soil to the evaporating surface; at the same time, the evaporation rate decreases gradually. This decrease corresponds to the first falling-rate stage. When the upward flow of water is less than evaporation rate E, drying of surface soil occurs before long and the evaporation rate decreases rapidly, which corresponds to the second falling-rate stage. The evaporation rate decreases with time due to the increased thickness of the drying zone, until a mass balance between the evaporation rate and the upward water flux is attained in the soil. When a drying zone is formed in the soil, the location of the evaporation surface moves down from the soil surface to the bottom of the drying zone. Within the drying zone, upward vapor flow becomes predominant. The water content of the drying zones in Figure 2.32 was about 0.04 cm^3 cm^{-3}.

B. Redistribution

After a halt in infiltration from the land surface, infiltrated water moves continuously downward in the soil. This downward flow is termed *redistribution* or, more suitably, *drainage stage*. Figure 2.33 shows the change in matric head profile during redistribution after a halt in infiltration (30). The matric head in the upper part of the wetted zone is high just after the halt in infiltration, but soon decreases. The matric head around the wetting front is low but increases gradually. The matric head gradient in the wetted zone is close to but larger than -1 cm cm^{-1}, while it is positive and large around the wetting front. Since the composition of matric head gradients and gravity are both driving forces of water flow, a small amount of downward flow in the wetted zone and a large amount of downward flow around the wetting front take place during redistribution.

Figure 2.34 shows soil moisture profiles during redistribution corresponding to the matric heads profiles given by Figure 2.33. The final soil moisture profile is an equilibrium distribution.

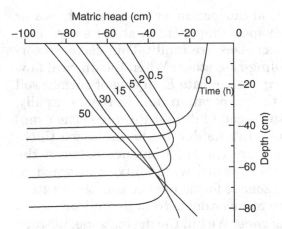

Figure 2.33 Gradual change of matric head profile during redistribution after a half in infiltration. (From Vachaud, G., and Thony, J.L., *Water Resour. Res.* 7(1):111–125 (1971). Modified in Nakano, M., *Transport Phenomena in Soils,* University of Tokyo Press, Tokyo (1991). With permission.)

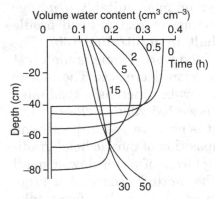

Figure 2.34 Gradual change of moisture profile corresponding to Figure 2.33. (From Vachaud, G., and Thony, J.L., *Water Resour. Res.* 7(1):111–125 (1971). Modified in Nakano, M., *Transport Phenomena in Soils,* University of Tokyo Press, Tokyo (1991). With permission.)

C. Water Flow in Deep Soils

1. Gravity-Predominant Flow (GPF)

Soils in deep zones are always wetted in the region where annual precipitation exceeds annual evapotranspiration and where precipitation is evenly distributed through the year. In regions where annual precipitation is less than annual evapotranspiration or in regions where dry and wet seasons are distinguished, the deep soil may either be dry or temporary wetted, according to the meteorological conditions.

In fields where the deep zones are always wetted, the water fluxes are not necessarily small, even though changes in the water content are small in deep zones. This flow of water in deep zones is characterized by a GPF in which the matric potential gradients are negligible compared with gravity. The Buckingham–Darcy equation for GPF is obtained by eliminating the matric potential gradient term $\partial \psi_m / \partial z$ from Equation (2.10) and by assuming that the unsaturated hydraulic conductivity K is isotropic ($K = K_z$), as is the propagation velocity

$$q_z = -K \tag{2.96}$$

and the continuity equation is given by

$$\frac{\partial \theta}{\partial t} = \frac{\partial K}{\partial z} \tag{2.97}$$

or, alternatively,

$$\frac{\partial \theta}{\partial t} + \zeta \frac{\partial \theta}{\partial z} = 0 \tag{2.98}$$

where

$$\zeta = -\frac{\partial K}{\partial \theta}$$

We can visually provide the propagation velocity in deep soils as illustrated in Figure 2.35. Assume that small moisture increase zones, shown as a very small wave (a), which may be generated by giving a small pulse water at a land surface,

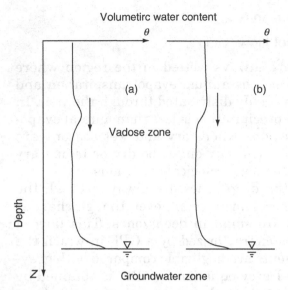

Figure 2.35 Gravity-predominant flows as a very small wave (a) and a very small step wave (b).

and as a very small step wave (b), which may be generated by a little increase of sprinkling intensity at a land surface, are both moving down through the deep soil at the velocity of $-dz/dt$. It is required that these moisture increase zones are so small that the term of matric potential gradient $\partial \psi_m / \partial z$ in the Buckingham–Darcy equation is negligible.

Since the right-hand side of Equation (2.98) is zero, the shapes of these waves are not transformed during the downward flows. If the right-hand side of Equation (2.98) is not zero, the shapes of any waves described by this equation will be transformed according to the function given in the right-hand side.

Under this circumstance, the downward propagation velocity $-dz/dt$ can characterize the GPF in wetted deep zones. This velocity is mathematically given by

$$\frac{dz}{dt} = -\frac{\partial K}{\partial \theta} \tag{2.99}$$

based on the method of characteristics as given in the appendix to this chapter.

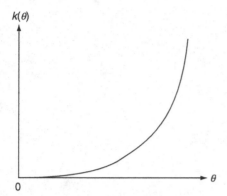

Figure 2.36 Unsaturated hydraulic conductivity function in a normal axis.

Figure 2.36 is a schematic of unsaturated hydraulic conductivity versus volumetric water content drawn in a plane of normal axes. Since the unsaturated hydraulic conductivity increases exponentially with water content, the gradient $\partial K / \partial \theta$ increases with water content, resulting in an increase in the absolute value of dz/dt with water content.

The mathematical features of Equation (2.99) explain the physical properties of flow in deep soils. The left-hand side of Equation (2.99) designates the propagation celerity of the location of a given θ. This means that whenever moisture profiles move in soils, the velocity of each part of the profile is different and hence the shape of the profile changes with time.

2. Wetting and Drainage in Deep Soils

When a wetting front moves down in a deep soil where a GPF exists, the shape of the wetting front may change, as illustrated in Figure 2.37, in which the greater the value of θ, the faster the location $z(\theta)$ moves down. Consequently, the shape of the wetting front will become sharper, which is mathematically identical to the result of a shock wave. On the other hand, when water is drained from soils, the shape of the moisture profile may change as illustrated in Figure 2.38,

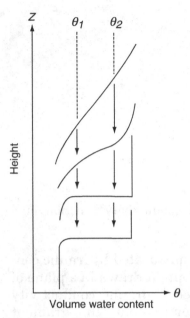

Figure 2.37 Moisture profile during infiltration in a deep soil. (After Shiozawa, S., Unsaturated soil water flow as a mechanism of groundwater recharge, PhD dissertation, The University of Tokyo (1988). With permission.)

where the profile is spread out more and more by the afore-mentioned characteristics of $z(\theta)$ and dz/dt.

Shiozawa and Nakano (31,32) investigated water flow in deep soils and simulated changes in matric head profiles and moisture profiles during a long-term drainage process by applying the GPF concept. Figure 2.39 shows the simulation where the groundwater level was at a depth of 20 m. The reason for the decrease in the gradient of matric heads during this drainage is mentioned above. It is noted again that the gradient of matric head $d\psi_m/dz$ exists in the deep soil zone but its value is negligibly small compared with 1.

Infiltrations of repeated but separate rains in a field are regarded to be rain pulses applied to the land surface. In this case water flows down by wetting the deeper soil and draining the upper soil because the water supply at the soil surface is tentative. When another, heavier rain is added separately

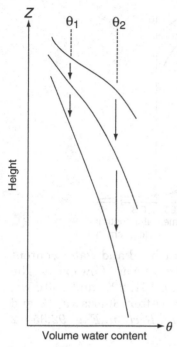

Figure 2.38 Moisture profile during drainage from a deep soil. (After Shiozawa, S., Unsaturated soil water flow as a mechanism of groundwater recharge, PhD dissertation, The University of Tokyo (1988). With permission.)

from the land surface, the new wetting front is superimposed on the former wetting front, due to the aforementioned features of $z(\theta)$ and dz/dt. Figure 2.40 shows the moisture profiles simulated by Shiozawa (32). Figure 2.40(a) is the profile 7 days after the first rain, and Figure 2.40(b) is the profile 8 days after the first rain and 1 day after the second rain, where the new wetting front is catching up with the earlier wetting front. Figure 2.40(c) is the profile 12 days after the first rain and 5 days after the second rain, where the old wetting front was superimposed by the new wetting front. Figure 2.40(d) is the profile 22 days after the first rain, when the superimposed wetting front reached the groundwater. This is the reason why the frequency of a rise in the groundwater level is less than the frequency of rain at

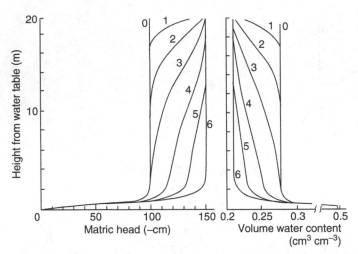

Figure 2.39 Simulated profiles of matric head and water content in a deep soil where gravity-predominant downward flow exists. The numbers 0 to 6 refer to 0.0, 6.3, 27.0, 65.0, 131, 223, and 530 days, respectively, after the start of drainage. (After Shiozawa, S. and Nakano, M., *Trans. Jpn. Soc. Irrig. Drain. Reclam. Eng.* 92:35–42 (1981). With permission.)

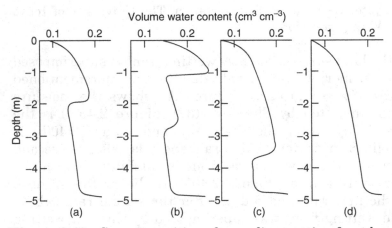

Figure 2.40 Superimposition of preceding wetting front by a new wetting front in gravity-predominant flow. (After Shiozawa, S., Unsaturated soil water flow as a mechanism of groundwater recharge, PhD dissertation, The University of Tokyo (1988). With permission.)

the land surface. The GPF concept proposed by Shiozawa (32) is thus an important and promising theory in the understanding of water flow in deep soils.

3. Moisture Profile Velocity and Mean Pore Water Velocity in Deep Soils

Anderson and Sevel (33) measured periodically tritium and soil moisture profiles in the unsaturated deep zone at a site covered with fine to medium sand and gravels in Denmark. The groundwater level was about 22 m and the mean precipitation was 780 mm per year. Figure 2.41 shows the monthly changes of soil moisture profiles where temporal water increasing zones are in black color. These zones, typically starting at October from the land surface, are moving down at the velocity about 3 to 3.5 m monthly. On the other hand, the mean pore water velocity, defined by the mean substantial water velocity in soils, obtained separately by measuring the downward movement of tritium concentration profiles in the same soil zone was 0.375 m per month.

Applying the GPF concept, the propagation velocity dz/dt of this soil moisture profile is thus estimated to be 3 to 3.5 m per month. On the other hand, the materials resolved in soil water are moving down at a velocity of about 0.375 m per month, since they are moving with substantial pore water. The relation between moisture profile velocity and mean pore water velocity in deep soils is significant in environmental issues such as soil and groundwater contamination.

The mean pore water velocity, v, taken positive upward, is defined by

$$v = \frac{q}{\theta} \tag{2.100}$$

where θ is the volumetric water content and q is the flux. Since the actual pore water velocity has a velocity distribution in each pore, v is regarded to be an averaged or apparent velocity.

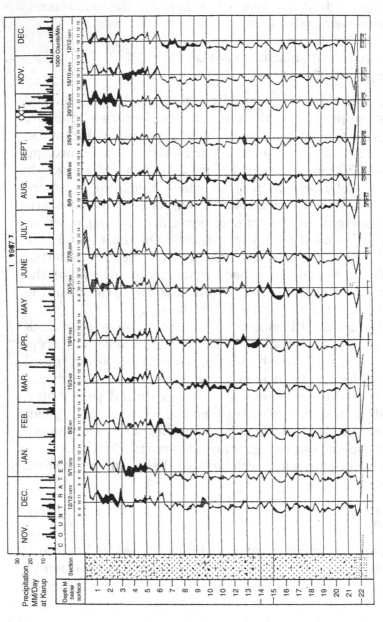

Figure 2.41 Soil moisture changes between successive soil moisture profiles from the unsaturated zone in Denmark. Black color is the zone where soil moisture increased. (After Anderson, L.J. and Sevel, T., Six Year's Environmental Tritium Profiles in the Unsaturated and Saturated Zones, Grønhøj, Denmark, *Isotope Techniques in Groundwater Hydrology, 1 IAEA-SM-182 / 1:3–20* (1974). With permission.)

When the flux of GPF is given by Equation (2.96), the mean pore water velocity is given by

$$v = -\frac{K}{\theta} \tag{2.101}$$

It should be noted that the propagation velocity ζ is the speed of shifting the moisture profile, while the mean pore water velocity is the speed of water itself.

If the unsaturated hydraulic conductivity is approximated by

$$K = a\theta^b \tag{2.102}$$

then the propagation velocity is given by

$$-\frac{\partial K}{\partial \theta} = -b\frac{K}{\theta} \tag{2.103}$$

Referring to Equation (2.101), the value of the right-hand side of Equation (2.103) is b times as high as that of the mean pore water velocity v. This means that in GPFs the profiles advance b times as fast as pore water advancement. The values of b for several natural soils are shown in Table 2.2 in which the values are ranged from 3.3 to 11.9. The larger the value of b, the larger the discrepancy between mean pore velocity and propagation velocity. This is why the downward advancement of isotopically labeled water is considerably behind that of the advancement of moisture profile in deep soils (33).

Table 2.2 Values of b for Several Soils

Soils	Saturated hydraulic conductivities (cm s^{-1})	b Values	References
Yolo light clay	7.4×10^{-6}	8.7	34
Silty clay loam	2.0×10^{-4}	7.7	34
Sand	1.7×10^{-2}	3.5	34
Sand	1.9×10^{-2}	3.3	35
Guelph loam	3.7×10^{-2}	11.9	35
Cecil Sandy loam	2.0×10^{-3}	10.1	35
Toyoura sand (washed)	1.4×10^{-2}	4.1	Unpublished
Clay loam	2.2×10^{-3}	9.1	Unpublished

Example   Explain the applicability of the GPF concept to the observed moisture profile velocity and the mean pore water velocity by Anderson and Sevel as shown in Figure 2.41.

The downward moisture profile velocity is 3 to 3.5 m per month, and the mean pore water velocity is 0.375 per month. The former is eight to nine times more than the latter. Taking into account that the texture of soil is fine to medium sand and referring to Table 2.2, this value reasonably agrees with value of b or a little larger than the values of b for sand. Assuming that the value of b for fine sand is close to that of sandy loam, the GPF concept seems to be acceptable for this actual case.

4.   Limitation of the Gravity-Predominant Flow Concept

The applicability of the GPF equation (2.97) is limited in the zones where the effects of matric head gradients are negligible. Water flow near land surfaces is always affected by rain infiltration, evaporation, and extraction of water by plant roots, all of which cause high values of matric head gradients near land surfaces. The matric head in soil near groundwater tends to equilibrate with the groundwater and hence has a matric head gradient of about -1, which is no longer negligible. Therefore, the GPF concept is not applicable to water flow in soils near the land surface and near groundwater.

APPENDIX: METHOD OF CHARACTERISTICS

Assume that a physical variable u is defined by position z and time t, and when $u(z, t)$ is determined by the first-order partial differential equation

$$\frac{\partial u}{\partial t} + \zeta \frac{\partial u}{\partial z} = f \tag{2.104}$$

where ζ and f are functions of z and t. The total differential of $u(z,t)$ is given by

$$du = \frac{\partial u}{\partial z} dz + \frac{\partial u}{\partial t} dt \tag{2.105}$$

and the differential of Equation (2.105) with t yields

$$\frac{\partial u}{\partial t} + \frac{\partial z}{\partial t}\frac{\partial u}{\partial z} = \frac{\partial u}{\partial t} \qquad (2.106)$$

Comparing Equations (2.104) and (2.105), it is recognized that the propagation celerity dz/dt is given by

$$\frac{dz}{dt} = \zeta \qquad (2.107)$$

Equation (2.107) is the characteristic curve on which a disturbance proceeds in the z–t plane. Integration of Equation (2.107) yields

$$z + B = \zeta(t + A) \qquad (2.108)$$

where A and B are constants of integration. The solution of Equation (2.104) is therefore equal to the solution of

$$\frac{du}{dt} = f \qquad (2.109)$$

on the characteristic curve (2.107). The solution of Equation (2.109) on Equation (2.107) is called a kinematic wave, and when $f = 0$, the wave proceeds in the velocity ζ without changing the shape of the wave.

REFERENCES

1. Buckingham, E. A., Studies on the movement of soil moisture, U.S. Dept. of Agric. Bureau of Soils, Bull. No. 38.
2. Nakano, M., *Transport Phenomena in Soils*, University of Tokyo Press, Tokyo (1991).
3. van Genuchten, M. Th., A closed-form equation for predicting the hydraulic conductivity of unsaturated soils, *Soil Sci. Soc. Am. J.*, 44:892–898 (1980).
4. Takagi, S., Analysis of the vertical downward flow of water through a two-layered soil, *Soil Sci.* 90:98–103 (1960).
5. Miyazaki, T., Kasubuchi, T., and Hasegawa, S., A statistical approach for predicting accuracies of soil properties measured by single, double and dual gamma beams, *J. Soil Sci.* 42: 127–137 (1991).

6. Srinilta, S., Nielsen, D. R., and Kirkham, D., Steady flow of water through two-layer soil, *Water Resour. Res.* 5(5):1053–1063 (1969).

7. Nakano, M. and Ichii, K., Measurement and prediction of hydraulic conductivity in unsaturated porous medium, *Trans. Jpn. Soc. Irrig. Drain. Reclam. Eng.* 69:29–34 (1977).

8. Richards, L. A., Capillary conduction of liquids in porous mediums, *Physics 1*: 318–333 (1931).

9. Miyazaki, T., Studies on the water retentivity and the permeability of "Masado," *Bull. Shikoku Natl. Agric. Exp. Stn.* 44: 186–199 (1984).

10. Tokunaga, K. and Sasaki, C., Actual water use, cropping pattern and irrigation planning in upland field areas, *J. Agric. Eng. Soc. Jpn.* 58(12):29–34 (1990).

11. van Genuchten, M. Th., Mass transfer studies in sorbing porous media. I. Analytical solutions, *Soil Sci. Am. J.* 40(4):473–480 (1976).

12. Miyazaki, T., Topography and movement of soil water, *Pedologist* 31(2):160–170 (1987).

13. Green, W. H. and Ampt, G. A., Studies on soil physics. 1. The flow of air and water through soils, *J. Agric. Sci.* 4(1):1–24 (1911).

14. Bouwer, H., Unsaturated flow in ground-water hydraulics, *J. Hydraul. Div. Am. Soc. Civil Eng.* 90(Hy5):121–144 (1964).

15. Mein, R. G. and Larson, C. L., Modeling infiltration during a steady rain, *Water Resour. Res.* 9(2):384–394 (1973).

16. Neuman, S. P., Wetting front pressure head in the infiltration model of Green and Ampt, *Water Resour. Res.* 12(3):564–566 (1976).

17. Swarzendruber, D. and Youngs, E. G., A comparison of physically-based infiltration equations, *Soil Sci.* 117(3):165–167 (1974).

18. Iwata, S., Tabuchi, T., and Warkentin, B. P., *Soil–Water Interactions*, Marcel Dekker, New York, pp. 257–323 (1995).

19. Philip, J. R., The theory of infiltration. 1. The infiltration equation and its solution, *Soil Sci.* 83:345–357 (1957).

20. Philip, J. R., Theory of infiltration. *Adv. Hydrosci.* 5:216–296 (1969).

21. Miyazaki, T., Nielsen, D. R., and MacIntyre, J. L., Early stage infiltration of water into horizontal and vertical soil columns, *Hilgardia* 52(6):1–24 (1984).

22. Gardner, W. R., Some steady state solutions of the unsaturated moisture flow equation with application to evaporation from a water table, *Soil Sci. 85*:228–232 (1958).
23. Hasegawa, S., Soil water movement in upland fields converted from paddy fields, *Soil Phys. Cond. Plant Growth Jpn. 53*:13–19 (1986).
24. Willis, W. O., Evaporation from layered soils in the presence of water table, *Soil Sci. Soc. Am. Proc. 24*:239–242 (1960).
25. Raats, P. A. C., Steady upward and downward flows in a class of unsaturated soils, *Soil Sci. 115*(6):409–413 (1973).
26. Warrick, A. W. and Yeh, T. C. J., One-dimensional, steady vertical flow in a layered soil profile. *Adv. Water Resour. 13*(4):207–210 (1991).
27. Warrick, A. W., Numerical approximations of Darcian flow through unsaturated soil, *Water Resour. Res. 27*(6):1215–1222 (1991).
28. Nakano, M., Miyazaki, T., and Maeda, S., Transport of soil moisture, salt and heat during evaporation, *Soil Phys. Cond. Plant Growth Jpn 58*:30–39 (1988).
29. Nakano, M., Soil water movement during the first stage of drying of a moist sandy soil under a very low drying rate, *Soil Sci. 124*(2):67–72 (1977).
30. Vachaud, G. and Thony, J. L., Hysteresis during infiltration and redistribution in a soil column at different initial water content, *Water Resour. Res. 7*(1):111–125 (1971).
31. Shiozawa, S. and Nakano, M., The mechanism of groundwater recharge as drainage process of unsaturated zone, *Trans. Jpn. Soc. Irrig. Drain. Reclam. Eng. 92*:35–42 (1981).
32. Shiozawa, S., Unsaturated soil water flow as a mechanism of groundwater recharge, PhD dissertation, The University of Tokyo (1988).
33. Anderson, L. J. and T. Sevel, Six year's environmental tritium profiles in the unsaturated and saturated zones, Crønhø, Denmark, *Isotope Techniques in Groundwater Hydrology, 1 IAEA-SM–182/1*:3–20 (1974).
34. Campbell, G. S., A simple method for determining unsaturated conductivity from moisture retention data. *Soil Sci. 117*(6):311–314 (1974).
35. Nishigaki, M., Some aspects on hydraulic parameters of saturated–unsaturated regional ground-water flow, *Soils Found. 23*(3):165–177 (1983).

3

Refraction of Water Flow in Soils

I. REFRACTION OF FLUX

Fluxes of water in soils change their magnitudes and directions depending on spatial variability of the hydraulic properties of soils. When hydraulic properties vary continuously with position in soils, the directions of streamlines may be curved. When hydraulic properties vary discontinuously with position in soils, the streamlines may be refracted at the boundaries between regions of different hydraulic properties. These curvings and refractions of streamlines occur not only in saturated soils but also in unsaturated soils. In layered soils, for example, both saturated and unsaturated hydraulic conductivities are usually different in each layer (1), resulting in refractions of fluxes of water at the boundaries of the layers.

Thus, refraction of fluxes influence several kinds of soil–hydrological processes: groundwater flow in heterogeneous soils (2,3), lateral flow of water during vertical percolation in layered soils (4), surface and subsurface water flow in slopes (5,6), and anisotropy of flow in saturated or unsaturated soils (6,7). As will be shown later, refraction of flux is not significant in one-dimensional water flow but is essential in two- or three-dimensional water flow in soils. In this chapter, two-dimensional refraction of water flow is described to provide a clear and simple example of refraction.

II. THEORY OF REFRACTION OF FLUX

von Kirchhoff (8) derived the refraction law of electric current crossing a boundary of different materials in a plane. Dachler (2) applied this law to groundwater flow and Raats (9,10) applied it to water flow in unsaturated soils. It is important to note that this refraction law of flux has nothing to do with the well-known refraction law of light, Snell's law, at the boundary of two materials of different refractive indexes.

Figure 3.1 shows the refraction of flux at the boundary of regions 1 and 2, respectively, having different hydraulic conductivities K_1 and K_2. The incidence angle of the incidence

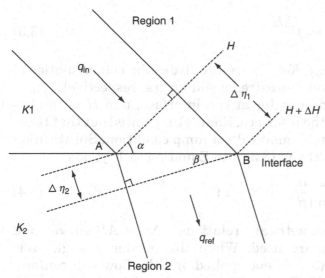

Figure 3.1 Refraction of flux of water at the interface of two regions.

flux q_{in} is α and the refraction angle of the refraction flux q_{ref} is β. The mass balance of the flux at the interface is deduced by equalizing the influx into plane AB and the outflux from plane AB on the interface. Denoting the hydraulic head at point A by H and that at point B by $H + \Delta H$, equihydraulic head lines are drawn normally to q_{in} in region 1 and to q_{ref} in region 2, as shown by dashed lines in Figure 3.1. The distance between these two equihydraulic head lines in region 1 is denoted as $\Delta \eta_1$ and that between those in region 2 is denoted as $\Delta \eta_2$. The mass balance equation is then given by

$$q_{\text{in}}AB \cos \alpha = q_{\text{ref}}AB \cos \beta \tag{3.1}$$

When α is zero, β must be zero based on this formula. This is the reason we deal with the vertical downward or upward water flow in horizontally layered soils without considering refraction.

Using Darcy's law, the fluxes q_{in} and q_{ref} are written as

$$q_{\text{in}} = -K_1(\psi_{\text{m}})\frac{\Delta H}{\Delta \eta_1} \tag{3.2}$$

$$q_{\text{ref}} = -K_2(\psi_m) \frac{\Delta H}{\Delta \eta_2} \tag{3.3}$$

where $K_1(\psi_m)$ and $K_2(\psi_m)$ are the hydraulic conductivities of the upstream and downstream soil layers, respectively, ψ_m is the matric head of water at the interface, and H is the hydraulic head at the interface. Raats (10) pointed out that these conditions are mathematically a jump condition. Substitution of Equations (3.2) and (3.3) into Equation (3.1) yields

$$\frac{K_1(\psi_m)}{K_2(\psi_m)} = \frac{\tan \alpha}{\tan \beta}, \quad 0 < \alpha \le 90° \tag{3.4}$$

where the geometrical relations $\Delta \eta_1 = AB \sin \alpha$ and $\Delta \eta_2 = AB \sin \beta$ are used. When the incidence angle α is zero, Equation (3.4) is not applied and the flow will conform to Equation (3.1). Total reflection, general in the refraction of light, does not occur in the refraction of flux, because the critical angle, which is defined as the least incidence angle to make the refraction angle more than 90°, is less than 90°.

In saturated soils, since the left-hand side of Equation (3.4) is constant, the refraction angle β is determined by the incidence angle α only. In unsaturated soils, on the other hand, since the left-hand side of Equation (3.4) is the function of the matric head ψ_m at the boundary, the refraction angle β is a function of both α and ψ_m. In the case of unsaturated soils, the left-hand side of Equation (3.4) ranges from very small to very large values, due to the rapid changes in the unsaturated hydraulic conductivities of soils with ψ_m. Figure 3.2 shows examples of unsaturated hydraulic conductivities of sand and fine beads (whose average particle size is 0.05 mm) as functions of ψ_m. Even though the saturated hydraulic conductivity of sand is larger than that of fine beads, the unsaturated hydraulic conductivity of fine beads exceeds that of sand when the matric head is less than -45 cm. If a flux of water crosses the contacted boundary from fine beads into sand, the refraction angle β may be larger than the incidence angle α when the boundary is so wetted that ψ_m is more than

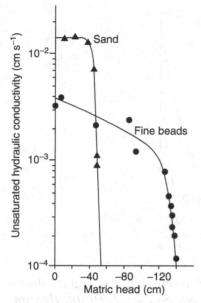

Figure 3.2 Unsaturated hydraulic conductivities of sand and fine beads. (After Miyazaki, T., *Soil Sci. 149*:317–319 (1990). With permission.)

−45 cm, but β may be smaller than α when the boundary is so dry that ψ_m is less than −45 cm.

Figure 3.3 shows the theoretical relation of refraction angles and incidence angles as functions of the parameter $K_1(\psi_m)/K_2(\psi_m)$, whose values are indicated in the figure. Note that when the value of K_1/K_2 is either very large or very small, extreme refraction may occur. For example, if an incidence angle α of flux is 30° at an interface where the value of K_1/K_2 is 0.01, the refraction angle β is given by

$$\beta = \tan^{-1}(100 \times \tan 30°) = 89.0° \qquad (3.5)$$

It is rather common to find such layers with interfaces between different types of soils, as exemplified here in fields.

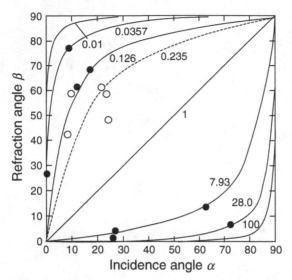

Figure 3.3 Theoretical relation of refraction angles and incidence angles. (After Miyazaki, T., *Trans. Jpn. Soc. Irrig. Drain. Reclam. Eng. 152*:75–82 (1991). With permission.)

III. JUMP CONDITION

Raats (9,10) showed that the refraction law of fluid at the interface between two porous media having different hydraulic properties is described mathematically in terms of jump conditions. In the vector field, shown in Figure 3.4, the flux crossing the interface at s is given by $\mathbf{q}(s)$ and its component in the direction normal to the interface is given by

$$q_n(s) = \mathbf{q}(s) \cdot \mathbf{n}(s) \tag{3.6}$$

where $\mathbf{n}(s)$ is the unit vector normal to the interface.

The jump condition is given by

$$q_n(s^-) - q_n(s^+) = 0 \tag{3.7}$$

where $q_n(s^-)$ and $q_n(s^+)$ denote the limits of q_n when the point s approaches from the upstream side to s and from the downstream side to s, respectively. Contrary, the nonhomogeneous jump condition is given by

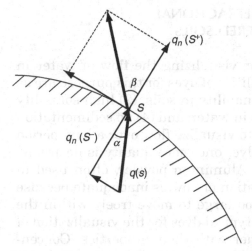

Figure 3.4 Jump condition on an interface s.

$$q_n(s^-) - q_n(s^+) = f \qquad (3.8)$$

where f is the source function on the interface (9). Except for special cases such as those of sources at the interface, we can apply the jump condition (Equation (3.7)) to water flow crossing the interface of two different soils.

Becker et al. (13) introduced the jump condition mentioned above to the variational boundary-value problems and gave a mathematically important suggestion: when we solve two-dimensional elliptic boundary-value problems numerically with finite element approximations in a domain consisting of a variety of materials, the jump terms at the interfaces do not require special attention. The reason that we may exclude the jump condition in the finite element method is described briefly in the appendix of this chapter.

On the other hand, in analytical solutions and finite difference methods, the jump condition is essential. In the analytical solution of flow equations, the jump condition at the interface between two domains must be regarded as a boundary condition of the given problem. In the finite difference method, it is preferred that the shapes of the interfaces be simple lines and that the nodes coincide with the interface.

IV. VISUALIZATION OF REFRACTIONAL WATER FLOW IN LAYERED SOILS

Several dyes are useful for visualizing the flow of water in soils (11). The required qualities of dyes for this purpose are as follows: (1) high distinguishability in soils, (2) high solubility in water, (3) low diffusivity in water, and (4) no sedimentation in water. When one needs to visualize flow over a long period without replenishing the dye, one more quality is desirable: slow dissolution in water. Aluminum powder, often used in hydraulics for the visualization of flow, is inadequate because the powder particles are too large to move freely within the soil pores. Table 3.1 shows typical dyes for the visualization of water flow in sand, together with their properties. Concentrated milk is distinguishable in dark soil but not in light-colored soils such as sand. Fluorescein satisfies criteria (2) to (4) but is not distinguishable in sand. Both potassium permanganate, most distinguishable but sedimenting somewhat in water, and methyl orange, preferable for quality but not distinguishable in light-colored soils, are recommended for visualization of flows in sand.

Figure 3.5(a) shows a special device for the visualization of flows in saturated soils, and Figure 3.5(b) illustrates the same device used for the visualization of flows in unsaturated soils. Horizontal flat flows from the left-hand side to the

Table 3.1 Typical Dyes to Visualize Water Flow in Sand and their Properties

Dye	Molecular formula	Molecular weight	Distinctness	Dispersivity	Remarks
Potassium permanganate	$KMnO_4$	158.0	Good	Small	Demands skill
Methyl orange	$C_{14}H_{14}N_3NaO_3S$	327.3	Medium	Very small	Demands care
Fluorescein	$C_{20}H_{12}O_6$	332.3	Poor	Small	Demands skill
Concentrated milk	—	—	Medium	Small	Easy to treat

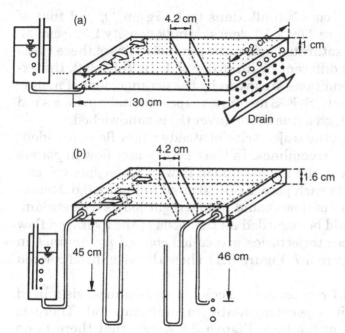

Figure 3.5 Device for the visualization of refractions of flow in (a) saturated and (b) unsaturated soils.

right-hand side of the container are generated by keeping constant heads at both the upstream and downstream boundaries. Table 3.2 gives the experimental conditions, where S represents sand; C, coarse beads; F, fine beads; and W, water. The saturated hydraulic conductivities of coarse beads (C) are $6.86 \times 10^{-1}\,\mathrm{cm\,s^{-1}}$ (bulk density $1.56\,\mathrm{g\,cm^{-3}}$), that of sand (S)

Table 3.2 Experimental Conditions for Visualization of Refractional Flow in Layer Model Soils

Flow type	Run number	Arrangement of layers	Angles of strips (°)	Anisotropy coefficients U
Saturated	1	S	—	0
	2	S–C–S	0	27
	3	S–C–S	25	6.9
	4	S–F–S	25	−0.87
Unsaturated	5	S–F–S	30	3.26

are 2.45×10^{-2}cm s^{-1} (bulk density 1.45 g cm^{-3}), and that of fine beads (F) are 3.09×10^{-3}cm s^{-1} (bulk density 1.52 g cm^{-3}). Note that the saturated hydraulic conductivities of these three materials are different by an order of 1. In Table 3.2, the experimental conditions are given by the arrangement of materials. For example, S–F–S means an experimental run for a sand layer (S) in which a fine beads layer (F) is sandwiched.

Generally, the trajectories of steady water flows are identified with the streamlines. In the case of water flow in porous media, however, the microscopical flow of water has velocity distributions in each pore, as illustrated in Figure 3.6. Hence, the trajectories of flow should not be regarded as true streamlines but should be regarded as the locus of the averaged flow of water. These trajectories are called *colored streamlines* in this book. Figure 3.7–Figure 3.11 show the visualized colored streamlines.

Figure 3.7 represents the colored streamlines, visualized using potassium permanganate, in uniform sand. There is no refraction in the flow. Figure 3.8 shows that there is no

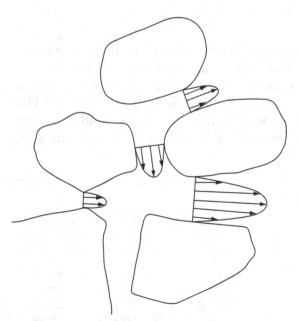

Figure 3.6   Microscopical velocity distributions in each soil pore.

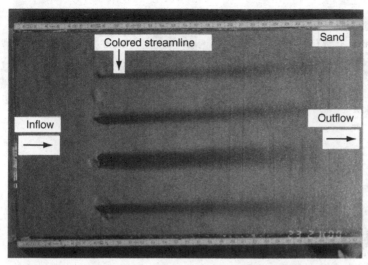

Figure 3.7 Colored streamlines in saturated uniform sand.

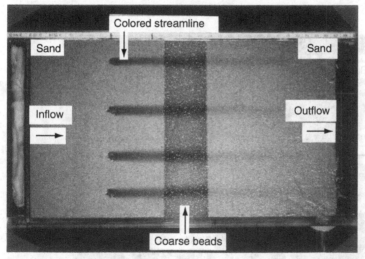

Figure 3.8 Colored streamlines in saturated sand–coarse beads–sand layer.

refraction in a sand–coarse beads–sand layer when the incidence angle is zero. Figure 3.9 shows that there are refractions at the boundaries of a sand–coarse beads–sand layer when the incidence angles are not zero. Figure 3.10 visualizes

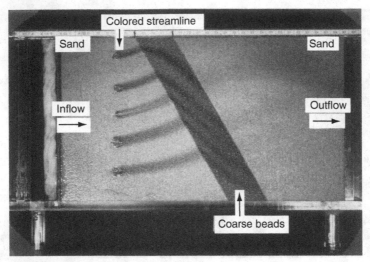

Figure 3.9 Colored streamlines in saturated sand–coarse beads–sand layer.

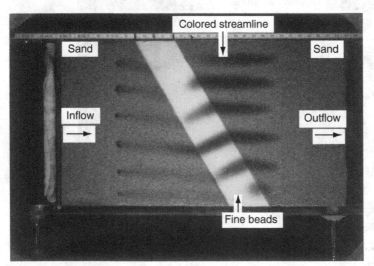

Figure 3.10 Colored streamlines in saturated sand–fine beads–sand layer.

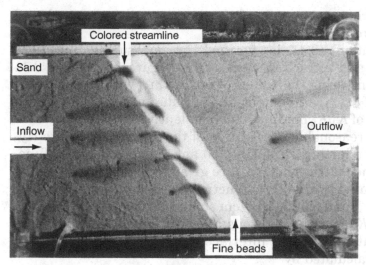

Figure 3.11 Colored streamlines in unsaturated sand–fine beads–sand layer.

the refraction of fluxes in a sand–fine beads–sand layer where a lower permeable layer (F) is sandwiched obliquely and the incidence angles are not zero. Figure 3.11 shows the refraction of fluxes in an unsaturated sand–fine beads–sand layer. By fixing the matric head to −45 cm at the left-hand side and −46 cm at the right-hand side, a steady unsaturated water flow was generated in the container. It is noted that the refraction angles in the saturated sand–fine beads–sand layer are smaller than the incidence angles (Figure 3.10), whereas the refraction angles in the unsaturated sand–fine beads–sand layer are larger than those of the incidence angles (Figure 3.11). This shows clearly the dependence of the refraction relations on the matric head (ψ_m) at the interface, as mentioned above.

V. VERIFICATION OF REFRACTION THEORY

A. Experimental Verification

The refraction law of water flux in soils has been taken for granted in classical groundwater flow theory and in the recent numerical analysis of two- or three-dimensional water flow, but experimental proofs of the law in both saturated and unsaturated soils are scarce.

Equation (3.4) is verified by measuring incidence angles α and refraction angles β in Figure 3.7–Figure 3.11 and by comparing the values of $\tan\alpha/\tan\beta$ with the values of $K_1(\psi_m)/K_2(\psi_m)$. Agreement of the value of $\tan\alpha/\tan\beta$ with the value of K_1/K_2 obtained independently at a given ψ_m value may give reasonable verification of the refraction theory. Hydraulic conductivities of sand (S) and fine beads (F) as functions of the matric head, ψ_m, are given in Figure 3.2. By using the saturated hydraulic conductivities of sand and fine beads, 2.45×10^{-2} and 3.09×10^{-3} cm s^{-1}, respectively, values of the ratio K_1/K_2 in saturated soils are determined easily. When a flux of water passes the interface from sand to fine beads, the value is calculated by

$$\frac{K_1(\psi_m = 0)}{K_2(\psi_m = 0)} = 7.93$$

and when a flux of water passes the interface from fine beads to sand, the value is calculated by

$$\frac{K_1(\psi_m = 0)}{K_2(\psi_m = 0)} = 0.126$$

Figure 3.10 shows two successive refractions; the first is the refraction at the interface from sand to fine beads and the second is that from fine beads to sand. At the first interface, the average value of α (incidence angle) is 29° and the average value of β (refraction angle) is 4°. The resultant value of $\tan\alpha/\tan\beta$ is 7.93. At the second interface, the averaged values of α and β are 4° and 29°, respectively, and the resultant value of $\tan\alpha/\tan\beta$ is 0.126. These two values are in excellent agreement with the values of K_1/K_2 given above. Thus, the observation given here verifies the use of refraction theory in the saturated layered soils.

Another example is shown in Figure 3.11, where five streamlines are visualized. The average values of α and β are 17.6° and 53.3°, respectively, and $\tan\alpha/\tan\beta$ is estimated to be 0.235. On the other hand, the value of K_1/K_2 is obtained from Figure 3.2 such that

$$\frac{K_1(\psi_m = -45.5 \text{ cm})}{K_2(\psi_m = -45.5 \text{ cm})} = 0.2 \text{ to } 0.6$$

Although the precise value of this quotient is hard to obtain due to the high sensitivity of the unsaturated hydraulic conductivity of sand to the matric head, the average value of $\tan \alpha / \tan \beta$ (0.235) is included in the range 0.2 to 0.6. This seems to be a reasonable verification of the use of refraction theory in the unsaturated layered soils.

B. Verification of a Numerical Method

The numerical method using two- and three-dimensional water flows in soils implicitly includes the refraction law, but again experimental verifications of such methods have been scarce. Therefore, the applicability of a numerical method of solving a flow equation implicitly, including the refraction law, is investigated here by comparing the solutions with the experimental results shown in Figure 3.7–Figure 3.11.

A two-dimensional horizontal steady flow of water in soils is governed by the partial differential equation

$$\frac{\partial}{\partial x}\left[K(x,y)\frac{\partial \psi_m}{\partial x}\right] + \frac{\partial}{\partial y}\left[K(x,y)\frac{\partial \psi_m}{\partial y}\right] = 0 \tag{3.9}$$

where ψ_m is the matric head and $K(x,y)$ is the hydraulic conductivity, a function determined by the position (x,y). The boundary-value problem of the flow in the region shown in Figure 3.12, where the length is L, the width is D, and the width of sandwiched strip is S, is solved numerically by applying Equation (3.9) with the boundary conditions (12)

$$\psi_m = \begin{cases} 0 \text{ cm}, & x = 0, \quad 0 < y < D \\ -1 \text{ cm}, & x = L, \quad 0 < y < D \end{cases} \tag{3.10}$$

$$\frac{\partial \psi_m}{\partial y} = 0, \quad 0 < x < L, \quad y = 0, \quad y = D$$

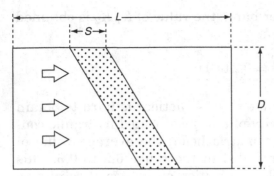

Figure 3.12   Sizes of the soil container.

for a saturated soil layer, and

$$\psi_m = \begin{cases} -45\,\text{cm}, & x = 0, \quad 0 < y < D \\ -45\,\text{cm}, & x = L, \quad 0 < y < D \end{cases} \tag{3.11}$$

$$\frac{\partial \psi_m}{\partial y} = 0, \quad 0 < x < L, \quad y = 0, \quad y = D$$

for an unsaturated soil layer.

The finite element method may be a suitable method for solving this boundary-value problem because the interfaces of the soil layer domains are oblique and, fortunately, use of this method does not require that we pay special attention to the jump conditions at these interfaces, as mentioned above (13). The theoretical background on this subject is given in the appendix of this chapter. Figure 3.13 shows the equipotential lines calculated every 0.05 cm with the traced streamlines in a sand–fine beads–sand layer (saturated), a sand–fine beads–sand layer (unsaturated), and a sand–water–sand layer (saturated). The streamlines obtained by the visualization experiments clearly intersect the calculated equipotential lines perpendicularly in all cases. This leads us to the conclusion that the numerical method demonstrated here is applicable to the particular case of refractional flow presented above.

Sand–fine beads–sand

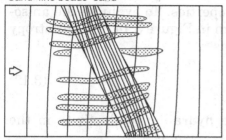

Sand–fine beads–sand

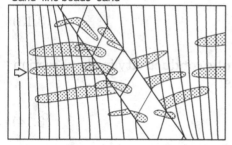

Sand–water–sand

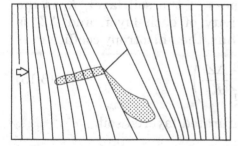

Figure 3.13 Numerical solutions of Laplace equation and visualized streamlines. (After Miyazaki, T., *Trans. Jpn. Soc. Irrig. Drain. Reclam. Eng. 152*:75–82 (1991). With permission.)

VI. REFRACTION AND ANISOTROPY COEFFICIENT

A. Anisotropy Coefficient in Saturated Layered Soils

In layered soils, the inclusive saturated hydraulic conductivity is anisotropic, even if each layer is isotropic. Alternatively, the anisotropy of saturated soils is often explained by a conceptual model composed of many thin parallel layers having different hydraulic properties. To evaluate the anisotropy of a layered soil as shown in Figure 3.14, the anisotropy coefficient, U, is defined (14) as

$$U = \frac{\langle K_\mathrm{h} \rangle}{\langle K_\mathrm{v} \rangle} \tag{3.12}$$

in which $\langle K_\mathrm{h} \rangle$ is the average hydraulic conductivity in the direction parallel to the layers, given as

$$\langle K_\mathrm{h} \rangle = \frac{1}{L} \sum_{i=1}^{i=n} K_i \Delta Z_i \tag{3.13}$$

and $\langle K_\mathrm{v} \rangle$ is the average hydraulic conductivity in the direction vertical to the layers, given as

$$\langle K_\mathrm{v} \rangle = L \left(\sum_{i=1}^{i=n} \frac{\Delta Z_i}{K_i} \right) \tag{3.14}$$

where L is the entire thickness of the soil layer, K_i is the saturated hydraulic conductivity of each layer, and ΔZ_i is the thickness of each layer. Thus, the anisotropy coefficient is

$$U = \frac{1}{L^2} \left(\sum_{i=1}^{i=n} K_i \Delta Z_i \right) \left(\sum_{i=1}^{i=n} \frac{\Delta Z_i}{K_i} \right) \tag{3.15}$$

and is constant for a given saturated layered soil.

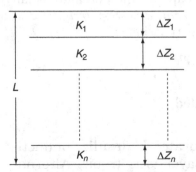

Figure 3.14   Schematic of layered soil.

B. Anisotropy Coefficient in Unsaturated Layered Soils

Unsaturated hydraulic conductivities in each layer are functions of matric head or water content. Hence, the anisotropy coefficient given in a formula such as Equation (3.12) is not constant but depends on the distribution of matric head or water content. Mualem (7) derived the anisotropy coefficient for unsaturated soils by integrating the unsaturated hydraulic conductivity of each layer as follows.

The average unsaturated hydraulic conductivity in the direction parallel to the layers is defined as

$$\langle K_{\mathrm{h}} \rangle = \frac{1}{L} \sum_{n=1}^{i=n} K_i(\psi_{\mathrm{m}}, K_{\mathrm{s}}) \Delta Z_l \tag{3.16}$$

and that in the direction normal to the layers is defined as

$$\langle K_{\mathrm{v}} \rangle = L \left[\sum_{i=1}^{i=n} \frac{\Delta Z_i}{K_i(\psi_{\mathrm{m}}, K_{\mathrm{s}})} \right]^{-1} \tag{3.17}$$

where $K_i(\psi_{\mathrm{m}}, K_{\mathrm{s}})$ is the unsaturated hydraulic conductivity, ψ_{m} is the matric head, and K_{s} is the saturated hydraulic conductivity of each layer. In layered soils, the saturated hydraulic conductivity K_{s} is constant in each layer, but the unsaturated hydraulic conductivity $K_i(\psi_{\mathrm{m}}, K_{\mathrm{s}})$ depends on both ψ_{m} and K_{s} in each layer. The anisotropy coefficient for unsaturated layered soils is then defined, in the same manner as Equation (3.15), as

$$U = \frac{1}{L^2} \left[\sum_{n=1}^{i=n} K_i(\psi_{\mathrm{m}}, K_{\mathrm{s}}) \Delta Z_i \right] \left[\sum_{n=1}^{i=n} \frac{\Delta Z_i}{K_i(\psi_{\mathrm{m}}, K_{\mathrm{s}})} \right] \tag{3.18}$$

In Equation (3.18), ψ_{m} and $K_i(\psi_{\mathrm{m}}, K_{\mathrm{s}})$ are regarded as constant in each layer.

Mualem (7) defined the anisotropy coefficient of unsaturated soils more generally as

$$U = \int K(\psi_{\mathrm{m}}, K_{\mathrm{s}}) f(K_{\mathrm{s}}) \, \mathrm{d}K_{\mathrm{s}} \int \frac{f(K_{\mathrm{s}})}{K(\psi_{\mathrm{m}}, K_{\mathrm{s}})} \mathrm{d}K_{\mathrm{s}} \tag{3.19}$$

where $f(K_s)$ is a density distribution function with the saturated hydraulic conductivity K_s.

C.   Refractive Anisotropy Coefficient

1.   Two-Layer Model

The anisotropy coefficients defined by Equation (3.15), (3.18), or (3.19) provide inclusive information on the anisotropy of layered soils. It is, however, preferable that the flux of water in each layer be related to the inclusive anisotropy of the given layered soil by using the anisotropy coefficient. Zaslavsky and Sinai (6) proposed a new definition of anisotropy coefficient in terms of the refraction law in a two-layered soil and predicted the generation of lateral water flow using the new anisotropy coefficient.

Figure 3.15 shows the two-layer model of refraction of fluxes at the interface, in which q_{in} is the incidence flux, q_1 is the flux in the second layer, and α is the incidence angle of q_{in}. The flux q_1 is decomposed into two components: one is the component q_z, parallel to q_{in}, and the other is the lateral component q_x, rectangular to q_{in}. It is evident that the lateral component q_x in the second layer is produced by the refraction of flow at the interface. The larger the ratio of q_x to q_z, the larger is the lateral component q_x for a given q_{in}. Hence,

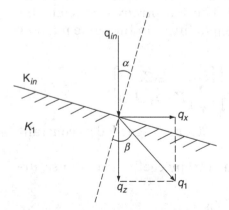

Figure 3.15   Two-layer model for the definition of refractive anisotropy coefficient.

the ratio of q_x to q_z may be a good index of the anisotropy of flow in two-layered soils.

In Figure 3.15, the ratio of q_x to q_z is given geometrically by

$$\frac{q_x}{q_z} = \tan(\beta - \alpha) \tag{3.20}$$

where β is the refraction angle. By applying the trigonometric relation

$$\tan(\beta - \alpha) = \frac{\tan\beta - \tan\alpha}{1 + \tan\beta\tan\alpha}$$

to the right-hand side of Equation (3.20) and using the refraction relation

$$\frac{K_{in}}{K_1} = \frac{\tan\alpha}{\tan\beta}$$

the left-hand side of Equation (3.20) is transformed to

$$\frac{q_x}{q_z} = \frac{(K_1/K_{in} - 1)\tan\alpha}{1 + (K_1/K_{in})\tan^2\alpha} \tag{3.21}$$

Defining the refractive anisotropy coefficient by

$$U_1^* = \frac{K_1}{K_{in}} - 1 \tag{3.22}$$

Equation (3.21) is simplified to

$$\frac{q_x}{q_z} = \frac{1}{2}\left(\frac{U_1^* \sin 2\alpha}{1 + U_1^* \sin^2\alpha}\right) \tag{3.23}$$

where elementary trigonometry is used.

Equation (3.23) along with Equation (3.22) provides the magnitude of the lateral flux component q_x generated by the refraction of a given influx q_{in} as a function of refractive anisotropy coefficient U_1^*. For example, if K_1 is equal to K_{in}, then U_1^* is 0 and the lateral flux q_x is also 0. If K_1 is much larger than K_{in}, U_1^* may also be extremely large, and hence the right-hand side of Equation (3.23) becomes

$$\frac{1}{2}\left(\frac{U_1^* \sin 2\alpha}{1 + U_1^* \sin^2 \alpha}\right) = \frac{1}{2}\left(\frac{\sin 2\alpha}{1/U_1^* + \sin^2 \alpha}\right) \rightarrow \cot \alpha \qquad (3.24)$$

This means that when K_1 is much larger than K_{in}, the ratio of q_x to q_z is given by

$$\frac{q_x}{q_z} \rightarrow \cot \alpha \qquad (3.25)$$

On the other hand, if K_1 is much smaller than K_{in}, then U_1^* approaches -1, and hence the right-hand side of Equation (3.23) becomes

$$\frac{1}{2}\left(\frac{U_1^* \sin 2\alpha}{1 + U_1^* \sin^2 \alpha}\right) \rightarrow -\tan \alpha \qquad (3.26)$$

This means that when K_1 is much smaller than K_{in}, the ratio of q_x to q_z is given by

$$\frac{q_x}{q_z} \rightarrow -\tan \alpha \qquad (3.27)$$

Thus, the refractive anisotropy coefficient U_1^* is useful for predicting how the incidence flux will refract at a given boundary of two-layer soils. Summarizing all the cases discussed above, the relation between the lateral component q_x and the refractive anisotropy coefficient U_1^* is given by

$$\begin{aligned} q_x &> 0, && \text{when } U_1^* > 0 \\ q_x &= 0, && \text{when } U_1^* = 0 \\ q_x &< 0, && \text{when } -1 < U_1^* < 0 \end{aligned} \qquad (3.28)$$

Figure 3.16 shows the physical meaning of Equation (3.28), where $q_x > 0$ means that the lateral flow along the interface is promoted by the refraction, $q_x = 0$ means that no refraction occurs at the interface, and $q_x < 0$ means that the lateral component of the incidence flow is diminished by the refraction. The larger the value of U_1^*, the more the lateral flow is generated, and the closer the value of U_1^* to -1, the more the lateral flow is diminished.

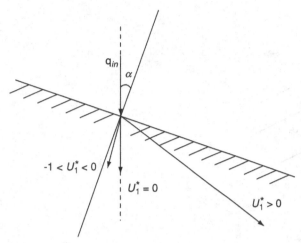

Figure 3.16 Dependence of lateral component q_x on U_1^*.

2. Multilayer Model

A multilayer model, expanded from the two-layer model of Zaslavsky and Sinai (6), is shown in Figure 3.17, where n is the number of layers, q_i ($i = 1, 2, \ldots, n$) are fluxes, β_i ($i = 1, 2, \ldots, n$) are refraction angles, and K_i ($i = 1, 2, \ldots, n$) are saturated or unsaturated hydraulic conductivities, functions of matric heads at the interfaces, in each layer. It is evident that each flux q_i is decomposed into components $q_{i,x}$ (perpendicular to the incidence flux q_{in}) and $q_{i,z}$ (parallel to the incidence flux q_{in}). Hence, the geometrical equation

$$\frac{q_{i,x}}{q_{i,x}} = \tan(\beta_i - \alpha) \tag{3.29}$$

is applied for any layer in the model. Equation (3.29) is developed using trigonometry in the same manner as in the two-layer model and by expanding the refraction theory as follows.

Assuming that streamlines within each layer do not refract or curve (this is achieved approximately by dividing a soil profile into many layers), Equation (3.4) is applied successively for this model, such that

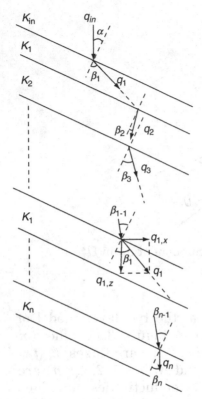

Figure 3.17 Multilayer model for the definition of refractive anisotropy coefficient.

$$\frac{K_{\text{in}}}{\tan \alpha} = \frac{K_1}{\tan \beta_1} = \frac{K_2}{\tan \beta_2} = \cdots = \frac{K_i}{\tan \beta_i} = \cdots$$

$$= \frac{K_n}{\tan \beta_n} \tag{3.30}$$

from which the relation

$$\tan \beta_i = \frac{K_i}{K_{\text{in}}} \tan \alpha \quad (i = 1, 2, \ldots, n) \tag{3.31}$$

is deduced. By substituting Equation (3.31) into the trigonometry form,

$$\tan(\beta_i - \alpha) = \frac{\tan \beta_i - \tan \alpha}{1 + \tan \beta_i \tan \alpha}$$

the left-hand side of Equation (3.29) is transformed into

$$\frac{q_{i,x}}{q_{i,z}} = \frac{(K_i/K_{in} - 1)\tan \alpha}{1 + (K_i/K_{in})\tan^2 \alpha} \tag{3.32}$$

Similar to the definition of anisotropy coefficient in the two-layer model, given by Equation (3.22), it may be reasonable to define the anisotropy coefficient of the ith layer in a multi-layer model by

$$U_i^* = \frac{K_i}{K_{in}} - 1 \tag{3.33}$$

Substitution of Equation (3.33) into Equation (3.32) yields

$$\frac{q_{i,x}}{q_{i,z}} = \frac{1}{2}\left(\frac{U_i^* \sin 2\alpha}{1 + U_i^* \sin^2 \alpha}\right) \tag{3.34}$$

Note that by the same discussion as that for the two-layer model, the range of the value of anisotropy coefficient is restricted to be

$$U_i^* \geq 1$$

The refractive anisotropy coefficient U_i^* provides a lateral component of flux relative to the incidence flux q_{in} in a given layer.

As discussed for the two-layer model, the larger the value of U_i^*, the larger the lateral component $q_{i,x}$. If K_i is much larger than K_{in}, the ratio of $q_{i,x}$ to $q_{i,z}$ in layer i of Figure 3.17 is given by

$$\frac{q_{i,x}}{q_{i,z}} \to \cot \alpha \tag{3.35}$$

On the other hand, if K_i is much smaller than K_{in}, the ratio of $q_{i,x}$ to $q_{i,z}$ is given by

$$\frac{q_{i,x}}{q_{i,z}} \to -\tan \alpha \tag{3.36}$$

When the saturated hydraulic conductivities of each layer in a layered soil decrease to the line of flow, the refractive anisotropy coefficients U_i^* are all negative. In this situation, saturated water flow will refract successively such that the refraction angles approach zero, as shown in Figure 3.18(a). But when the saturated hydraulic conductivities of each soil layer increase in the line of flow, the refractive anisotropy coefficients U_i^* are all positive, and therefore the refraction angles will gradually increase, as shown in Figure 3.18(b). This refraction in saturated layered soils is presumably seen in fields with layered soils.

Since the unsaturated conductivities of soils generally change markedly with changes in matric heads, the refractive anisotropy coefficients U_i^* defined by Equation (3.33) will also change markedly in unsaturated layered soils, resulting in considerable change in the refraction angles at interfaces with changes in matric heads.

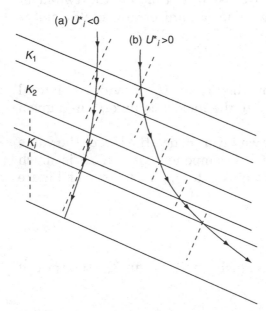

Figure 3.18 Successive refractions of flux in layered soils with (a) negative and (b) positive U_i^* values.

APPENDIX: THE JUMP CONDITION AND THE FINITE ELEMENT METHOD

The jump condition is another mathematical formulation of the conservation law at the interface between two regions. The finite element method, which uses variational concepts to construct an approximation of the solution over a collection of finite elements, is favorable for the simulation of water flow in heterogeneous soils, since the jump conditions at the interfaces of plural domains are always satisfied theoretically in this method.

Becker et al. (13) showed how the jump conditions are satisfied in the variational boundary-value problem of such domains, as shown in Figure 3.19, where the problem is designated by the sum of domain Ω and boundary $\partial\Omega$. Domain Ω consists of two domains, Ω_1 and Ω_2, interfacing at Γ,

$$\Omega = \Omega_1 + \Omega_2 \tag{3.37}$$

The boundary $\partial\Omega$ consists of the fundamental boundary condition, denoted as $\partial\Omega_1$, and the natural boundary condition,

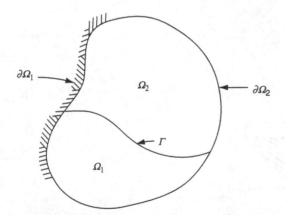

Figure 3.19 Domain of two-dimensional boundary-value problem. (From Becker, E.B., Carey, G.F., and Oden, J.T., *Finite Elements: An Introduction*, Vol. I, Prentice-Hall, Englewood Cliffs, NJ (1981), with several deletions. With permission.)

denoted as $\partial\Omega_2$, both of which are designated independently on domains Ω_1 and Ω_1; thus,

$$\partial\Omega = \partial\Omega_1 + \partial\Omega_2 \tag{3.38}$$

The entire boundary $\partial\Omega$ is also denoted by subtracting the interface Γ from both boundaries $\partial(\Omega_1)$ and $\partial(\Omega_2)$, which as shown in Figure 3.20, surround domains Ω_1 and Ω_2, respectively, as

$$\partial\Omega = [\partial(\Omega_1) - \Gamma] + [\partial(\Omega_2) - \Gamma] \tag{3.39}$$

Let us consider the two-dimensional steady flow equation

$$\frac{\partial}{\partial x}\left[k(x,y)\frac{\partial h'}{\partial x}\right] + \frac{\partial}{\partial x}\left[k(x,y)\frac{\partial h'}{\partial y}\right] = 0 \tag{3.40}$$

where $h'(x,y)$ is the suction, a smooth function in Ω, and $k(x,y)$ is the hydraulic conductivity defined in each domain, discon-

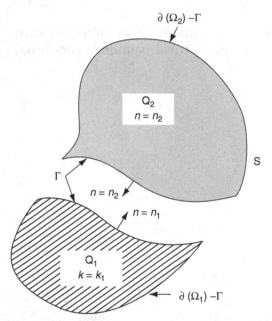

Figure 3.20 Decomposition of regions of integration for boundary integrals around $\partial(\Omega_1)$ and $\partial(\Omega_2)$. (After Becker, E.B., Carey, G.F., and Oden, J.T., *Finite Elements: An Introduction*,Vol. I, Prentice-Hall, Englewood Cliffs, NJ (1981). With permission.)

tinuous at the interface Γ. In the finite element method, $h'(x,y)$ is approximated by a linear function $h(x,y)$,

$$h(x,y) = \sum_{j=1}^{j=N} h_j \phi_j(x,y) \tag{3.41}$$

where h_j are values of h at each node (x_j, y_j) of the finite elements and $\phi_j(x,y)$ are basis functions satisfying

$$\phi_i(x_j, y_j) = \begin{cases} 1 & \text{if } i = j \\ 0 & \text{if } i \neq j \end{cases} \tag{3.42}$$

When $h'(x,y)$ is approximated by $h(x,y)$, the residual r of Equation (3.40),

$$r = \frac{\partial}{\partial x}\left[k(x,y)\frac{\partial h}{\partial x}\right] + \frac{\partial}{\partial y}\left[k(x,y)\frac{\partial h}{\partial y}\right] \tag{3.43}$$

is not zero. The approximation $h(x,y)$ is determined in such a way that the integral of r over Ω, weighted with a test function v, is set equal to zero such that

$$\int_\Omega rv \, d\Omega = 0 \tag{3.44}$$

Integration of Equation (3.44) must be conducted separately over Ω_1 and Ω_2 since the second derivatives of h are not integratable along the interface Γ. This procedure gives

$$\int_{\Omega_1}\left[\frac{\partial}{\partial x}\left(k\frac{\partial h}{\partial x}\right) + \frac{\partial}{\partial y}\left(k\frac{\partial h}{\partial y}\right)\right]v \, dx \, dy$$
$$+ \int_{\Omega_2}\left[\frac{\partial}{\partial x}\left(k\frac{\partial h}{\partial x}\right) + \frac{\partial}{\partial y}\left(k\frac{\partial h}{\partial y}\right)\right]v \, dx \, dy = 0 \tag{3.45}$$

where each integral corresponds to each domain. By using the Gauss divergence theorem, the first term of Equation (3.45) is divided again into two terms as

$$\int_{\Omega_1} \left[\frac{\partial}{\partial x}\left(k\frac{\partial h}{\partial x} \right) + \frac{\partial}{\partial y}\left(k\frac{\partial h}{\partial y} \right) \right] v \, dx \, dy$$

$$= -\int_{\Omega_1} k\left(\frac{\partial h}{\partial x}\frac{\partial v}{\partial x} + \frac{\partial h}{\partial y}\frac{\partial v}{\partial y} \right) dx \, dy + \int_{\partial(\Omega_1)} vk\frac{\partial h}{\partial n} \, ds \quad (3.46)$$

and the second term of Equation (3.45) is divided as

$$\int_{\Omega_2} \left[\frac{\partial}{\partial x}\left(k\frac{\partial h}{\partial x} \right) + \frac{\partial}{\partial y}\left(k\frac{\partial h}{\partial y} \right) \right] v \, dx \, dy$$

$$= -\int_{\Omega_2} k\left(\frac{\partial h}{\partial x}\frac{\partial v}{\partial x} + \frac{\partial h}{\partial y}\frac{\partial v}{\partial y} \right) dx \, dy + \int_{\partial(\Omega_2)} vk\frac{\partial h}{\partial n} \, ds \quad (3.47)$$

Noting that

$$\partial(\Omega_1) = [\partial(\Omega_1) - \Gamma] + \Gamma$$
$$\partial(\Omega_2) = [\partial(\Omega_2) - \Gamma] + \Gamma$$

the second terms on the right-hand sides of Equations (3.46) and (3.47) are rewritten, respectively, as

$$\int_{\partial(\Omega_1)} vk\frac{\partial h}{\partial n} \, ds = \int_{\partial(\Omega_1)-\Gamma} vk\frac{\partial h}{\partial n} \, ds + \int_\Gamma \left(vk\frac{\partial h}{\partial n} \right)_1 ds \quad (3.48)$$

$$\int_{\partial(\Omega_2)} vk\frac{\partial h}{\partial n} \, ds = \int_{\partial(\Omega_2)-\Gamma} vk\frac{\partial h}{\partial n} \, ds + \int_\Gamma \left(vk\frac{\partial h}{\partial n} \right)_2 ds \quad (3.49)$$

Adding Equations (3.48) and (3.49), we obtain

$$\int_{\partial(\Omega_1)} vk\frac{\partial h}{\partial n ds} + \int_{\partial(\Omega_2)} vk\frac{\partial h}{\partial n} ds$$

$$= \left[\int_{\partial(\Omega_1)-\Gamma} vk\frac{\partial h}{\partial n} \, ds + \int_{\partial(\Omega_2)-\Gamma} vk\frac{\partial h}{\partial n} \, ds \right] \quad (3.50)$$

$$+ \left[\int_\Gamma \left(vk\frac{\partial h}{\partial n} \right)_1 ds + \int_\Gamma \left(vk\frac{\partial h}{\partial n} \right)_2 ds \right]$$

The sum in the first set of brackets on the right-hand side of Equation (3.50) is equal to

$$\int_{\partial\Omega} vk\frac{\partial h}{\partial n} \, ds \quad (3.51)$$

and the sum in the second set of brackets on the right-hand side of Equation (3.50) is equal to

$$\int_{\Gamma} \left[\left(k \frac{\partial h}{\partial n} \right)_1 + \left(k \frac{\partial h}{\partial n} \right)_2 \right] v \, ds \qquad (3.52)$$

where the notation

$$\left(k \frac{\partial h}{\partial n} \right)_i$$

indicates that the term inside the parenthesis is evaluated on region i. Since the unit vector $\mathbf{n}$ is the outward normal to a given region, $\mathbf{n}_1$ at each point on Γ is equal to the negative of $\mathbf{n}_2$ and Equation (3.52) is rewritten as

$$\int_{\Gamma} \left[\left(k \frac{\partial h}{\partial n} \right)_{\Gamma^-} - \left(k \frac{\partial h}{\partial n} \right)_{\Gamma^+} \right] v \, ds = \int_{\Gamma} [q_n(\Gamma^-) - q_n(\Gamma^+)] v \, ds$$

$$(3.53)$$

According to the jump condition given in Equation (3.7), this integral is zero.

Returning to Equations (3.46) and (3.47) and adding the first terms on the right-hand sides, we obtain

$$-\int_{\Omega_1} k \left(\frac{\partial h}{\partial x} \frac{\partial v}{\partial x} + \frac{\partial h}{\partial y} \frac{\partial v}{\partial y} \right) dx \, dy - \int_{\Omega_2} k \left(\frac{\partial h}{\partial x} \frac{\partial v}{\partial x} + \frac{\partial h}{\partial y} \frac{\partial v}{\partial y} \right) dx \, dy$$

$$= -\int_{\Omega} k \left(\frac{\partial h}{\partial x} \frac{\partial v}{\partial x} + \frac{\partial h}{\partial y} \frac{\partial v}{\partial y} \right) dx \, dy \qquad (3.54)$$

Finally, the integration Equation (3.44) is rewritten as

$$\int_{\Omega} k \left(\frac{\partial h}{\partial x} \frac{\partial v}{\partial x} + \frac{\partial h}{\partial y} \frac{\partial v}{\partial y} \right) dx \, dy - \int_{\partial \Omega} v k \frac{\partial h}{\partial n} \, ds = 0 \qquad (3.55)$$

In Galerkin's method, a well-known approximation in finite element method, the basis function ϕ_i ($i = 1, 2, \ldots, N$) is used as a test function v. Substituting Equation (3.41) into Equation (3.55) and writing ϕ_i in place of v, Equation (3.55) is rewritten in the form

$$\sum_{j=1}^{j=N} \left[\int_{\Omega} k \left(\frac{\partial \phi_i}{\partial x} \frac{\partial \phi_i}{\partial x} + \frac{\partial \phi_i}{\partial y} \frac{\partial \phi_i}{\partial y} \right) \right] h_j \, dx \, dy = \int_{\partial \Omega} \phi_i k \frac{\partial h}{\partial n} \, ds$$

$$(3.56)$$

or following custom,

Kh = **F**

where the components of **K**, **h**, and **F** are given by

$$K_{ij} = \int_{\Omega} k \left(\frac{\partial \phi_i}{\partial x} \frac{\partial \phi_i}{\partial x} + \frac{\partial \phi_i}{\partial y} \frac{\partial \phi_i}{\partial y} \right) dx \, dy$$

$$h_j = h(x_j, y_j) \qquad\qquad\qquad (3.57)$$

$$F_i = \int_{\partial \Omega} \phi_i k \frac{\partial h}{\partial n} ds$$

This formulation is just the same using the finite element approximation for one domain Ω, where no interface exists in the domain. Hence, we can use this approximation for domains constructed of two or more subdomains having interfaces inside the whole domains. Becker et al. (13) noted, however, that the location of nodes and element boundaries should coincide as closely as possible with interfaces at which jumps in the coefficient k occur.

REFERENCES

1. Miller, D. E. and Gardner, W. H., Water infiltration into layered soil, *Soil Sci. Soc. Am. Proc.* 26:115–119 (1962).
2. Dachler, R., *Grundwasserströmung*, Springer-Verlag, Vienna, p. 141 (1936).
3. Bear, J., *Hydraulics of Groundwater*, McGraw-Hill, New York, pp. 100–102 (1979).
4. Miyazaki, T., Water flow in unsaturated soil in layered slopes, *J. Hydrol.* 102:201–214 (1988).
5. Zaslavsky, D. and Sinai, G., Surface hydrology. III. Causes of lateral flow, *J. Hydraul. Div. ASCE, 107 HY1, Proc. Paper 15960*:37–52 (1981).

6. Zaslavsky, D. and Sinai, G., Surface hydrology. IV. Flow in sloping, layered soil, *J. Hydraul. Div. ASCE, 107 HY1, Proc. Paper 15961:*53–64 (1981).
7. Mualem, Y., Anisotropy of unsaturated soils, *Soil Sci. Soc. Am. J. 48:*505–509 (1984).
8. von Kirchhoff, S., Uber den Durchgang eines elektrischen Stromes durch eine Ebene, insbesondere durch eine Kreisförmige, *Ann. Phys. Chem. 64*(4):497–514 (1845).
9. Raats, P. A. C., Jump condition in the hydrodynamics of porous media, *Joint IAHR and ISSS Symposium on Transport Phenomena in Porous Media*, Guelph, Ontario, Canada, pp. 155–173 (1972).
10. Raats, P. A. C., Refraction of a fluid at an interface between two anisotropic porous media, *J. Appl. Math. Phys. 24:*43–53 (1973).
11. Miyazaki, T., Visualization of refraction water flow in layered soils, *Soil Sci. 149:*317–319 (1990).
12. Miyazaki, T., Refraction of steady water flow in layered soils, *Trans. Jpn. Soc. Irrig. Drain. Reclam. Eng. 152:*75–82 (1991).
13. Becker, E. B., Carey, G. F., and Oden, J. T., *Finite Elements: An Introduction*, Vol. I, Prentice-Hall, Englewood Cliffs, NJ, pp. 131–174 (1981).
14. Hird, C. C. and Humphreys, J. D., An experimental scheme for the disposal of micaceous residues from the china clay industry, *Q. J. Eng. Geol. 10:*177–194 (1977).

4

Preferential Flow

I. CLASSIFICATION OF PREFERENTIAL FLOW

It is widely recognized that preferential flows of water in soils have a great effect on hydrological processes in fields. In addition, the role of preferential water flow involving chemicals or contaminations is so important in the environment that many symposia have focused on this topic (1). However, the definition and classification of preferential flow are still under investigation. The classification of preferential flows based on their phenomenological features as proposed by Kung (2,3) is useful. Taking account of his classification, we classify preferential flows in soils into three types: bypassing flow, fingering flow, and funneled flow. Figure 4.1 is a schematic diagram of these flows.

Bypassing flow, denoted by B in Figure 4.1(a), is local flow in such highly permeable zones as a macropore (right-hand side) and a crack (left-hand side) in heterogeneous soil. This flow takes place either when the macropores and cracks are open to the atmosphere or when the water pressure within the macropores and cracks is positive. When, for example, ponded water infiltrates a soil matrix whose macropores are open to the land surface, water will infiltrate the macropores (flux B) faster than the soil matrix (flux I), due to bypassing flows, and sooner or later, the bypassing flows will infiltrate the soil matrix across the walls of the macropores. If macropores exist beneath a layer of groundwater, water will preferentially flow within the macropores. Under certain conditions, this type of flow will occasionally be transformed into pipe flows. Readers are directed to a paper by Beven and Germann (4) in which many types of bypassing flows are reviewed.

On the other hand, when ponded water infiltrates a soil matrix within which macropores or cracks are buried, and when the soil matrix is under suction, water will not infiltrate the buried macropores but will infiltrate only the soil matrix. In this case, macropores are not preferential for the flow of water but are obstacles to such flow. Further details on the flow of water in buried macropores are given in Chapter 9.

Fingering flow, denoted by Fi in Figure 4.1(b), is a partial flow that looks like fingers. This flow takes place mainly in

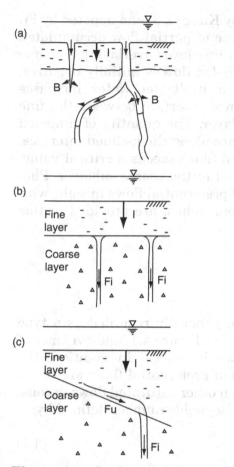

Figure 4.1 Preferential flows denoting infiltration (I): (a) bypassing flow (B); (b) fingering flow (Fi); (c) funneled flow (Fu).

relatively coarse layers overlaid with finer soils. It has been shown that under some conditions, fingering flows are also generated even in uniform soils (5,6). Fingering flows occur under both ponded and sprinkled water at the land surface. Fingering flows have been analyzed by applying Darcy's law, but the physical characteristic of this flow remains to be studied, because the Reynolds number of a fingering flow is much larger than that of flow in a soil matrix, and hence the applicability of Darcy's law has to be investigated.

Funneled flow, so called by Kung (2,3) and denoted by Fu in Figure 4.1(c), is another type of partial flow accumulated along the inclined bottom of a fine layer overlaying a coarse sublayer. The pressure of funneled flow is initially negative, and due to the accumulation of infiltrated water, increases with distance along the inclined interface between the fine top layer and the coarse sublayer. The quantity of funneled flow also increases with distance along the inclined interface. When the pressure of funneled flow exceeds a critical value, fingering flow will be generated in the coarse sublayer. Phenomenological explanations of preferential flows in soils, with the help of some modeling approaches, are provided in this chapter.

II. GENERAL FEATURES OF PREFERENTIAL FLOW

Preferential flows in soils are generally regarded as a type of saturated flow, as illustrated in Figure 4.1, and evidence of unsaturated preferential flows is scarce. In addition, the velocities and local quantities of preferential flows are considered to be large compared with other saturated flows in soils. Table 4.1 shows the values of Reynolds number defined by

$$Re = \frac{Du}{v} \tag{4.1}$$

Table 4.1 Reynolds Numbers for Several Types of Flow

Type of flow	Re
Water flow in clay	10^{-9}
Motion of a sperm	10^{-5}
Blood flow in a capillary	10^{-3}
Water flow in sand	10^{-3}–10^{-2}
Preferential flow in soil	1–10
Rise of an air bubble in beer	30
Speed of a bullet train	10^{7}

where D is the characteristic length (m), u is the average velocity of the flow (m s^{-1}), and v is the coefficient of kinematic viscosity of the fluid (m^2 s^{-1}). The Reynolds number depends on the choice of the characteristic length of the flow in question. In the case of water flow in soils, when the average diameter of soil particles is used as the characteristic length, laminar flow is predominant when the Reynolds number is less than about 10, and, consequently, Darcy's law is applicable to the flow. The inertia force increases with the increase in Reynolds number, and turbulent flow is generated when the Reynolds number exceeds this value. When the Reynolds number exceeds about 150 to 300, the turbulent flow becomes predominant. Hence, the range of application of Darcy's law in soils is limited to flows where Reynolds numbers are less than about 10.

The Reynolds number of preferential flows in soils is estimated to be less than 10 but is very close to this value. For example, the velocity of vertical-down fingering flow in coarse glass particles, whose average diameter is 1 mm, is about 1 cm s^{-1}. Since the coefficient of kinematic viscosity of water is close to 0.01 cm^2 s^{-1}, the Reynolds number for this fingering flow is estimated from Equation (4.1) to be 10. Since Reynolds numbers over 10 are possible depending on the physical condition of the flow in soils, the physical characteristics of preferential flows should be investigated individually.

III. BYPASSING FLOW

A. Vertical Bypassing Flow

There is much evidence of the existence of bypassing flow in fields. van Stiphout et al. (7) investigated the bypassing flow in an undisturbed soil column taken from a grassland field. Abundant cracks and worm holes were observed in the topsoil at a depth of 0 to 60 cm, while the subsoil at a depth of 60 to 135 cm contained only worm holes. Before and after the application of water from the land surface, they measured the suction profiles and compared them with profiles simulated numerically. The same procedures were performed twice.

There were very large discrepancies between the measured and simulated profiles, as shown in Figure 4.2.

The initial profile of $\log|h|$, where h is the suction of water in the soil, ranged from 1.5 to 2.7 cm in the topsoil, which is 120 cm deep. Investigating the measured profiles after the first and second applications of water (open and

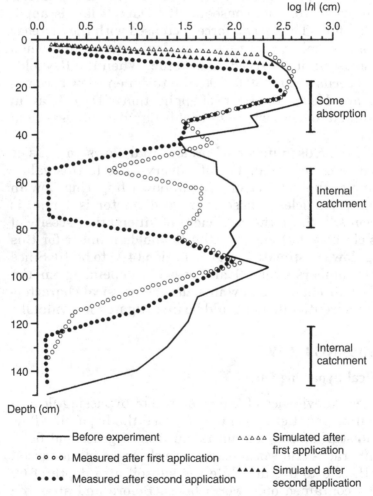

Figure 4.2 Measured and simulated suction profiles in an undisturbed soil column. (After van Stiphout, T. P. J., van Lanen, H. A. J., Boersma, O. H., and Bouma, J., *J. Hydrol.* **95**:1–11 (1987). With permission.)

solid circles, respectively, in Figure 4.2), remarkable decreases in suction (i.e., increase in water content) appeared at a depth of 50 to 80 cm, especially after the second application. If the flow of infiltrated water follows Darcy's law, the suction profiles will be similar to the calculated profiles shown by triangles in Figure 4.2, but the measured profiles were far from the calculated profiles. After investigating the horizontal cross sections of the soil, they concluded that bypassing flow within cracks and worm holes was predominant in the topsoil at a depth of 0 to 60 cm and the accumulation of bypassing flow brought about a quick decrease in suction in the topsoil layer. The other decrease in suction (i.e., increase in water content), which appeared at a depth of 110 to 140 cm, was due to the bypassing flow and its accumulation in the deep zone, where there were worm holes but no cracks. It is surprising that the macropores are channeled to the land surface and open to the atmosphere in such a deep zone.

Thus, they concluded that the ability to predict suction profiles (triangles) based on Darcy's law is very poor in these heterogeneous soils. Underestimation of the hydraulic conductivity of the soil might be one reason for these discrepancies. In addition, we doubt that Darcy's law is applicable to this vertical preferential flow, as the large macropore size causes an increase in the Reynolds number. Assuming that the sizes of cracks and worm holes in the soil are greater than 1 mm and that the vertical velocity of water flow in them is greater than 1 cm s^{-1}, the Reynolds number of this vertical bypassing flow is estimated to be above 10, which results in deviation from the range of applicability of Darcy's law.

B. Lateral Bypassing Flow

When hydraulic head gradients exist in the horizontal direction in soils with macropores, lateral bypassing flows may take place. A typical lateral bypassing flow is observed in rice paddies equipped with an underdrain. It is well known that even in a paddy of heavy clay soil, an underdrain is fairly effective because lateral bypassing flow in cracks is predominant in such fields.

Inoue et al. (8) investigated the hydraulic properties of soils in rice paddies developed in cracks. They prepared a large lysimeter 30 by 70 m wide and 0.7 m deep with a thick vinyl sheet at the bottom (Figure 4.3), packed alluvial clayey subsoil in the lysimeter, and cultivated it for 10 years, yielding rice or soybeans. At the center of the lysimeter a main underdrain pipe was buried and mole drains were perforated at a depth of 0.35 m every 1.2 m in the direction normal to the main drain, as shown in Figure 4.3. After 10 years of cultivation, an undisturbed test soil block 5 m long, 1.44 m wide, and 0.7 m deep was formed by excavating around the test block. The lateral water flow in this test soil block was generated by using different water levels in the pits at both short sides of the block, as shown in Figure 4.4. Along the longer sides the block was sealed by vinyl sheets and the gaps between the block and original field along the longer sides were refilled with soil so that no space remained. Since the lysimeter had been cultivated continuously, the average saturated hydraulic conductivity of topsoil at a depth of 0 to 12 cm was around 10^{-3} cm s^{-1}, with an average bulk density of 1.15 g cm^{-3}. The average saturated hydraulic conductivity of subsoil at a depth of 12 to 70 cm was about 10^{-6} cm s^{-1}, with an average bulk density of 1.46 g cm^{-3}.

White paint dissolved in water was poured into the left short-side pit and the water level was maintained constant by

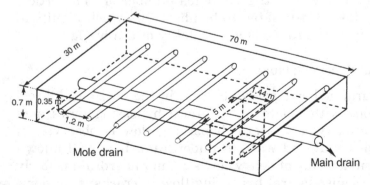

Figure 4.3 Large lysimeter with drains and an undisturbed test soil.

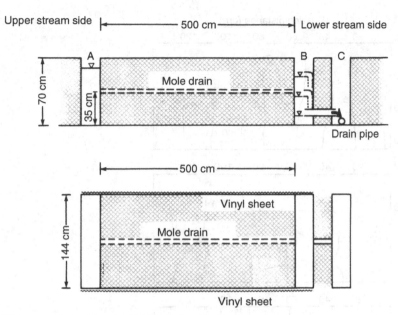

Figure 4.4 Test soil block in the lysimeter.

adding the white solution. The water-dissolved paint flowed in the test soil block, dyeing the flow paths, and drained into the right short-side pit, denoted as B in Figure 4.4. The rate of drainage was measured at the adjacent pit C, into which water overflowed at a given level in pit B and ran through a pipe. Figure 4.5 shows the cracks dyed by white paint, as seen by the naked eye, in cross sections at distances of 15, 115, and 200 cm from the upper stream side of the test soil block. The solid lines are dyed cracks, the dashed lines are undyed cracks, and the dotted areas are dyed walls of cracks. It is evident that water-dissolved paint did not flow in the soil matrix but flowed mainly in the network of cracks and partly in the mole drain. This is a direct verification of bypassing flow in the field.

The physical property of this lateral bypassing flow was investigated by analyzing the flow data. Assuming that Darcy's law is applicable to the lateral flow of water in the test soil block, the flux q is given by

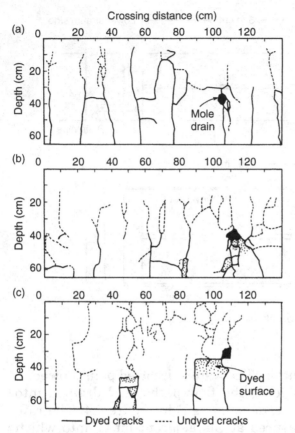

Figure 4.5 Dyed cracks by white paint in cross sections at distances of 15, 115, and 200 cm from the upper pit. (After Inoue, H., Hasegawa, S., and Miyazaki, T., *Trans. Jpn. Jpn. Soc. Irrig. Drain. Reclam. Eng. 134*:51–59 (1988). With permission.)

$$q = -K\frac{\Delta H}{L} \tag{4.2}$$

where ΔH is the difference in water level at both sides A and B in Figure 4.4, L is the horizontal distance of the test soil block, and K is the saturated hydraulic conductivity. The value of q is obtained experimentally by

$$q = \frac{Q}{\langle A \rangle} \tag{4.3}$$

where Q is the total amount of lateral water flow through the soil block per unit time and $\langle A \rangle$ is the average sectional area perpendicular to flow in the soil block.

Figure 4.6 shows the relation between Q and ΔH at various water levels in the pits. If Darcy's law is applicable to the flow, the relation between Q and ΔH will be linear and Q will be zero when ΔH is zero. The rate of increase of Q with ΔH decreased in curve 1, where water levels in both pits were higher than the mole drain. The nonlinearity of curve 1 was attributed to a high Reynolds number of flow in the mole drain (about 44). The quantity Q increased linearly in curve 2, where the water level was higher in the upstream pit and

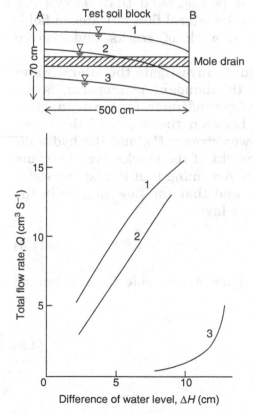

Figure 4.6 Relation between Q and ΔH.

lower in the downstream pit. The quantity Q was relatively small and the rate of increase of Q with ΔH was fairly non-linear in curve 3, where the water levels in both pits were lower than the level of the mole drain. The nonlinearity of curve 3 was attributed to the geometric features of the bottom of cracks where the pathways were disturbed by narrow crack walls and clogged by lumps.

Apparently, curve 2 in Figure 4.6 satisfies Darcy's law, but curves 1 and 3 deviated from the law. According to the calculation by Inoue et al. (8), due to the low values of the flow velocities, the Reynolds numbers of these lateral bypassing flows were from 1 to 3, which is small enough to apply Darcy's law. The hydraulic conductivity of the bypassing flow, designated by curve 2, is estimated to be 0.15 cm s^{-1}. Although this value is fairly high, it is suggested that Darcy's law is restrictively applicable to the lateral flow of water in fields with a highly developed network of cracks under given conditions.

Another method, useful to investigate the lateral water flow across the section with abundant macropores, is the analysis of surface slopes of groundwater. Figure 4.7(a) gives conceptually the relation between the depth of the upper stream side H_1 and the lower stream H_2, and the hydraulic head h measured by the height of the cracks dyed by white paint, as seen in Figure 4.5. Assuming that the lateral water flow is under steady state and that the flow mainly in the cracks is described in Darcy's law

$$Q = -Kh \frac{dh}{dx} \tag{4.4}$$

the integration from the upper stream side to an arbitrary lateral distance x

$$Q \int_0^x dx = -K \int_{H_1}^h h \, dh \tag{4.5}$$

results in

$$\frac{Q}{K} = -\frac{1}{x}\left(\frac{h^2 - H_1^2}{2}\right) \tag{4.6}$$

Since the saturated hydraulic conductivity K is easily obtained by substituting h for H_2,

$$K = \frac{2QL}{H_1^2 - H_1^2} \tag{4.7}$$

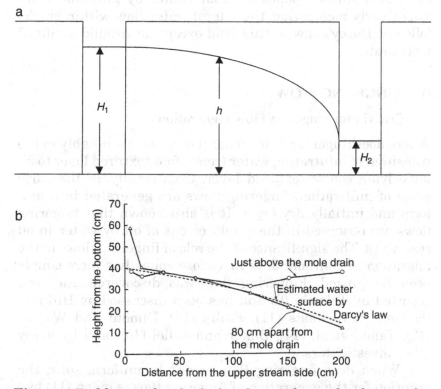

Figure 4.7 Measured and calculated groundwater levels in cracks under a lateral steady flow. (After Inoue, H., Hasegawa, S., and Miyazaki, T., *Trans. Jpn. Jpn. Soc. Irrig. Drain. Reclam. Eng.* *134*:51–59 (1988). With permission.)

and the left-hand side of Equation (4.6) is determined by

$$\frac{Q}{K} = \frac{H_1^2 - H_2^2}{2L} \tag{4.8}$$

The resultant relation between h and x is given by

$$h = \sqrt{H_1^2 - \frac{x}{L}\left(H_1^2 - H_2^2\right)} \tag{4.9}$$

Figure 4.7(b) shows the height of cracks dyed by white paint along the stream distance very close to the mole drain zone and 80 cm away from the mole drain zone, obtained by Inoue et al. (8). The latter distribution was in agreement with the calculated surface slopes of groundwater by Equation (4.9). This clearly means that the lateral water flow within cracks followed Darcy's law in this field except for around artificial mole drain.

IV. FINGERING FLOW

A. Criteria for Fingering Flow Generation

Where does fingering flow occur? It occurs most notably in the transition of infiltrating water from a fine-textured layer to an underlying coarse-textured layer. Occasionally, in the early stage of infiltration, fingering flows are generated in a uniform and initially dry layer. It is also known that fingering flows are observed in the displacement of oil by water in oil reservoirs. The significance of the role of fingering flows in the migration of contaminants in vadose zones has increasingly been recognized. Fingering flow was discovered and first reported by Tabuchi (9) and has been discussed by Hill and Parlange (10), Raats (11), Philip (12), Diment and Watson (13), Tamai et al. (14), Baker and Hillel (15), and by many other investigators.

   When does fingering flow occur? In uniform soils, the criterion for the generation of fingering flow is given (11) by

$$h_0 > h_c \tag{4.10}$$

where h_0 is the suction at the soil surface and h_c is the suction at the wetting front. Since the suction at the wetting front is affected by both water content and air pressure ahead of the wetting front, the compression of air has an influence on the generation of fingering flow. If ponded water infiltrates a uniform soil without compressing the air ahead of the wetting front, the soil surface becomes wetter than the wetting front and thus h_0 is smaller than h_c, resulting in no fingering flow. On the other hand, if the infiltration is accompanied by air compression ahead of the wetting front, the value of h_c decreases with time, resulting in the generation of fingering flow. The instability of the wetting front in uniform soils has been investigated by several researchers (9, 11, 12).

In layered soils, the criterion for the generation of fingering flow is given by

$$K_u(h_w) > q_t \tag{4.11}$$

where $K_u(h_w)$ is the hydraulic conductivity of the sublayer at its water entry suction value h_w and q_t is the flux through the top layer. This criterion is often satisfied in a layered soil consisting of a coarse soil overlaid by a fine soil, and fingering flows have been investigated in such soils by many of the researchers noted above. Note that the value of $K_u(h_w)$ is approximated by the saturated hydraulic conductivity K_s, even though the latter is a little larger than the former (15).

B. Hydrodynamics of Fingering Flow

The physical mechanism leading to the various types of finger flows in soils are still under study and, especially, their lateral hydrodynamics are vague. Kawamoto and Miyazaki (16) investigated suction changes during infiltrations of simulated raindrop water into Toyoura sand, whose particles are larger than 0.105 mm and smaller than 0.21 mm, packed in an acrylic board container with inside dimensions of 50 cm width, 50 cm height, and 1 cm thickness. Table 4.2 shows all the runs corresponding to the initial water contents and artificial rainfall intensities. Figure 4.8 is the traced wetting fronts classified into three types: low-swell fingers (Runs 1 to

Table 4.2 Experimental Conditions

	Rainfall intensity (I_R)		
Initial water content (w_i)	15 mm h^{-1}	30 mm h^{-1}	180 mm h^{-1}
Air dry (0.0%)	Run 1	Run 2	Run 3
0.5%	Run 4	Run 5	Run 6
1.0%	Run 7	Run 8	Run 9

Source: Kawamoto, K. and Miyazaki, T., *Soils Found. 39*(4):79–91 (1999). With permission.

4), high-swell finger (Run 5), and wavy to plane fronts (Runs 6 to 9).

Figure 4.9(a) shows the suction change during the passing of a low-swell finger tip at depths of 7.5 and 22.5 cm in Run 2. When a finger tip arrives at a designated depth, the suction at this point instantaneously decreases and, after reaching the lowest value, it rapidly increases and finally remains a steady value. Since the suction of the wetted zone including the surface zone by this finger is, from Figure 4.9, estimated to be about 34 cm, and since the suction at the wetting front of this finger is estimated to be less than 34 cm, criterion (4.4) is well confirmed.

Figure 4.9(b), on the other hand, shows the suction change of Run 9 during the passing of a plane front infiltration at a depth of 22.5 cm. When the plane wetting front arrives at 22.5 cm depth, the suction at this point gradually increases to a final steady value. There is no indication of fingers and, as guessed earlier, criterion (4.4) is not applicable.

The persistence of finger flows may strongly depend on the lateral water movement within the finger. Kawamoto and Miyazaki (16) showed that the suction changes with swelling of both low-swell finger (Figure 4.10) and high-swell finger and pointed out that the lateral suction gradients in low-swell fingers are, against our presumption, much higher than those in high-swell fingers. Based on this finding, they concluded that the lateral water movements with fingering flows are dominated mainly by the unsaturated hydraulic conductivity and that the large value of lateral suction gradients does not directly dominate the horizontal water velocity from the center of fingers to the surrounding dry sand.

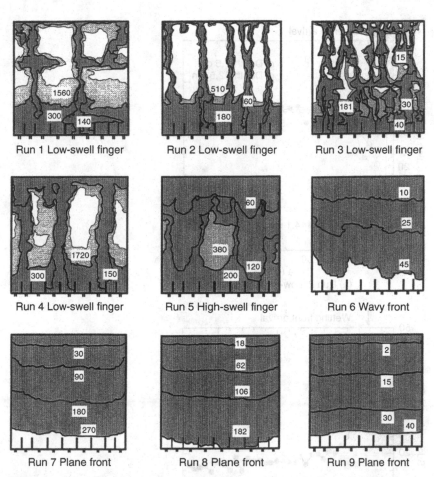

Figure 4.8 Tracings of wetting fronts (the numbers in diagrams refer to the time after rainfall [min]). (After Kawamoto, K. and Miyazaki, T., *Soils Found. 39*(4):79–91 (1999). With permission.)

V. FUNNELED FLOW

A. Nature of Funneled Flow

Kung (2) investigated water flow in a huge soil block of inter-bedding coarse sand layers in a potato field with a ridged surface. Using a red dye, he was able to distinguish an accumulation of water above the interfaces between the upper fine sand and lower coarse sand, and subsequent lateral spreading

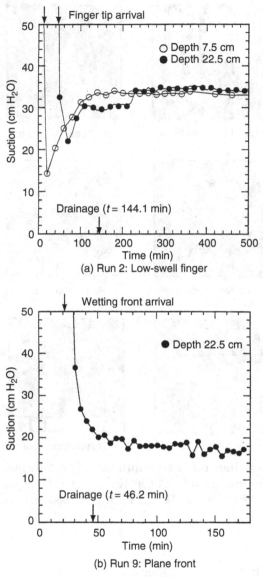

Figure 4.9 Suction changes with wetting fronts passage. (After Kawamoto, K. and Miyazaki, T., *Soils Found. 39*(4):79–91 (1999). With permission.)

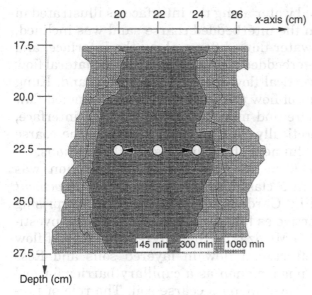

Run 4: Low-swell finger

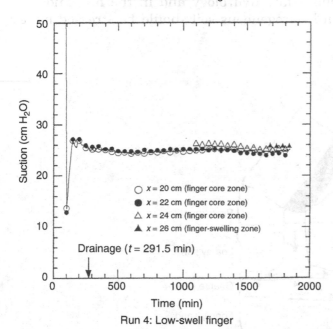

Run 4: Low-swell finger

Figure 4.10 Suction changes with swelling of a low-swell finger. (After Kawamoto, K. and Miyazaki, T., *Soils Found.* 39(4):79–91 (1999). With permission.)

flow of accumulated water along the interface as illustrated in Figure 4.11. When the interbedded coarse sand was inclined, the accumulated water flowed down along the interface, and at the top of the interbedded coarse sand layer, the lateral flow changed back to vertical flow into the lower fine sand. Kung (3) named this type of flow *funneled flow*. Since funneled flow is accumulated more and more along the inclined interface, he predicted theoretically that fingering flow into the coarse layer would follow funneled flow under special conditions.

As noted by Kung (3), the funnel phenomenon was first documented in a classic film entitled *Water Movement in Soil*, produced by Gardner in 1960 (17). It is surprising that flow as distinctive as funneled flow had not been investigated in detail until Miyazaki (18) analyzed this type of flow as an extreme refractional flow in layered soils and Ross (19) analyzed the phenomenon as a capillary barrier formed by fine-grained soil overlaying a coarse soil. The role of funneled flow in subsurface hydrology and in the migration of contaminants in heterogeneous soil should be stressed more strongly.

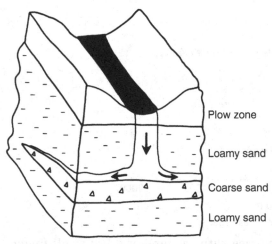

Figure 4.11 Funneled flow in a field. (After Kung, K.-J. S., *Geoderma* 46:51–58 (1990). With permission.)

B. Experimental Observation

Miyazaki (20–22) conducted an experimental investigation of funneled flow and subsequent fingering flow using Hele–Shaw models. Figure 4.12 shows an experimental device composed of an acrylic board container of 2 m width, 1 m height, and 0.03 m thickness, a rain simulator, a drain pipe, and a measurement system. A small acrylic board container of 0.5 m width, 0.7 m height, and 0.02 m thickness was also used in the experiment. Within these boxes, sand whose water content was 0.02 g g^{-1} was packed over coarse 1-mm glass beads spread uniformly. The sand just above the interface was packed carefully so as not to disturb the interface. The interface was formed horizontally or at an angle. Bulk densities of sands ranged from 1.39 to 1.45 g cm^{-3} and those of glass beads ranged from 1.45 to 1.46 g cm^{-3}. The rain intensity was controlled by changing the elevation of a constant-head water reservoir (Mariotte bottle). Table 4.3 shows the experimental conditions and some of the results.

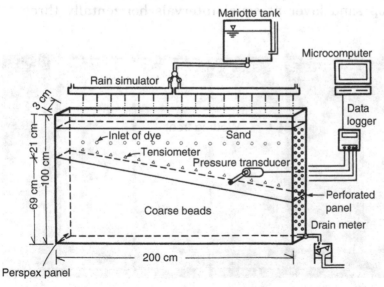

Figure 4.12 Hele–Shaw model to generate funneled flow.

Table 4.3　Experimental Conditions and Principal Results

Run number	Container size	Angle of interface (°)	Rain intensity (mm h⁻¹)	Number of fingers	Number of fingers per length (m⁻¹)	Average thickness of fingers (mm)
1	Small	0	100	4	8.0	5.3
2	Large	0	140	18	9.0	—
3	Large	0	120	18	9.0	—
4	Small	5	100	4	8.0	6.9
5	Large	10.9	50	7	3.5	6.0
6	Large	10.9	100	16	8.0	5.5
7	Large	13.5	150	15	7.5	—
8	Small	15	100	3	6.0	8.8
9	Small	30	100	1	2.0	11.3

Source: Miyazaki, T., Diversion capacity of capillary barrier in a layered slope, in *Trends in Hydrology*, S. G. Pandalai, Ed., Council of Scientific Information, Trivandrum, pp.469–475 (1994). With permission. Miyazaki, T., *Trans. Jpn. Soc. Irrig. Drain. Reclam. Eng. 179*:49–56 (1995). With permission.

　　　　Figure 4.13–Figure 4.15 show the behavior of the wetting front and subsequent fingering flow during infiltration in the larger container. Before the start of infiltration, clumps of a dye (potassium permanganate powder) were embedded in the top sand layer at equal intervals horizontally through

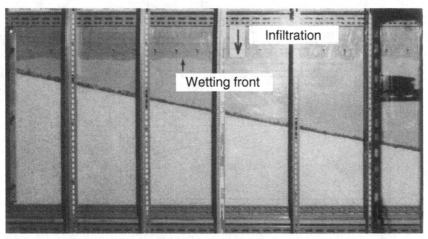

Figure 4.13　Early stage of infiltration.

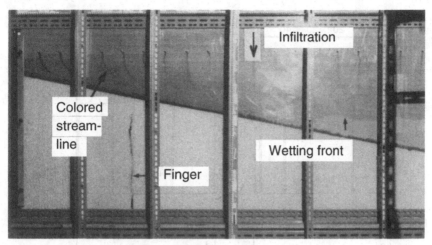

Figure 4.14 Cessation of wetting front advancement at the interface.

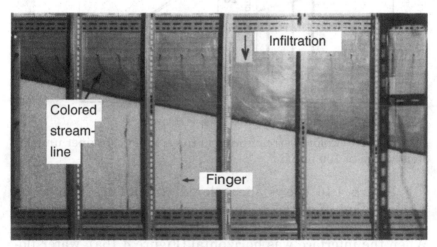

Figure 4.15 Final stage of infiltration.

small holes on the front panels. As water infiltrated downward, the horizontal wetting front in the top sand layer remained stable while the colored streamlines curved along the inclined interface, which suggested the existence of funneled flow.

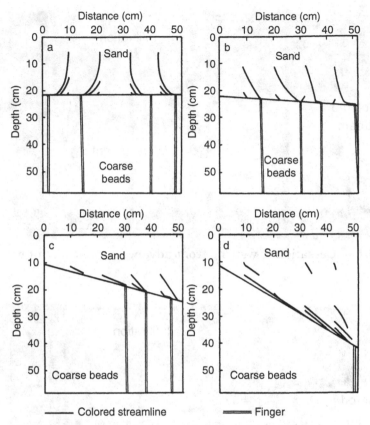

Figure 4.16 Colored streamlines in the smaller container under a steady rain of 100 mm h^{-1}.

Figure 4.16 shows the colored streamlines formed by a steady flow of water in layered soil models in the smaller container under a constant rain intensity of 100 mm h^{-1}. When the interface was horizontal (Figure 4.16a), water flowed laterally above the interface and four fingers appeared. When the slope of the interface was 5°, water in the sand layer showed a tendency to flow down along the slope of the interface (Figure 4.16b). When the slope of the interface was 15°, funneled flow arose evidently along the interface and three fingers appeared (Figure 4.16c). When the slope of the interface was 30°, funneled flow and, subsequently, one finger were generated (Figure 4.16d).

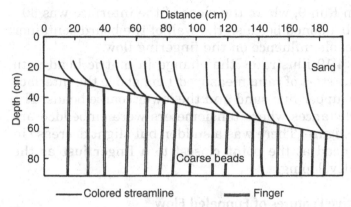

Figure 4.17 Colored streamlines in the larger container under a steady rain of 100 mm h^{-1}.

Figure 4.17 shows the colored streamlines of a steady flow of water in the top sand layer and subsequent fingering flows in the second, coarse bead layer generated in the larger container under a constant rain intensity of 100 mm h^{-1}. The interface was inclined 10.9° and 16 fingers appeared in the coarse bead layer. Figure 4.18 shows the case when the rain intensity was 50 mm h^{-1}, when seven fingers appeared in the coarse layer.

The numbers of fingers in each experiment and their average thicknesses are given in Table 4.3. The thickness of

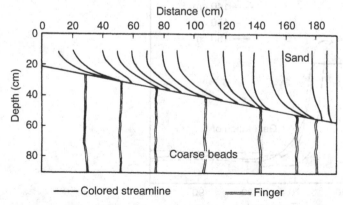

Figure 4.18 Colored streamlines in the larger container under a steady rain of 50 mm h^{-1}.

the finger in Run 9, where the slope of the interface was 30°, should be abandoned because the wall of the board container had a noticeable influence on the fingering flow.

Figure 4.19 illustrates the change in matric head with time from the start of rain, measured 1 cm above the interface between the upper fine sand and the lower coarse beads. The horizontal distances where tensiometers were embedded are given in the figure. There was a sudden but slight decrease in the matric head at the point closest to a finger just as the finger was developing.

C. Qualitative Features of Funneled Flow

All the experimental results mentioned above demonstrate the features of funneled flow and subsequent fingering flow in layered soils. Figure 4.20 shows a conceptual funneled flow and subsequent fingering flows in a layered soil where a coarser layer is overlaid by a finer layer. L_i $(i = 1, 2, \ldots, n)$ denotes the distance between the fingers numbered $i - 1$ and i, l_i $(i = 1, 2, \ldots, n)$ denotes the width of the area in the upper fine sand layer that controls the flux of the finger numbered i, and n is the total number of fingers.

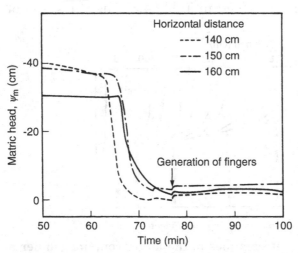

Figure 4.19 Change of matric heads 1 cm above the interface.

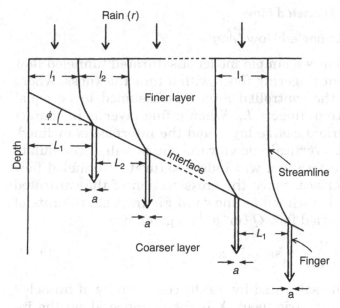

Figure 4.20 Conceptual funneled flow and subsequent fingering flows in layered soil.

There are five conspicuous features in these flows:

1. The horizontal distance L_1, the distance between the left sidewall and the location where the first finger appears, is proportional to the inclination ϕ.

2. The average value $\langle L \rangle$, defined by

$$\langle L \rangle = \frac{L_2 + L_3 + \cdots + l_n}{n - 1} \tag{4.12}$$

    does not depend on the inclination ϕ.

3. The average value $\langle l \rangle$, defined by

$$\langle l \rangle = \frac{l_2 + l_3 + \cdots + l_n}{n} \tag{4.13}$$

    does not depend on the inclination ϕ.

4. The average value $\langle L \rangle$ is inversely proportional to the intensity of precipitation r.

5. Finger thickness does not depend on the inclination ϕ.

D. Modeling Funneled Flow

1. Saturated Funneled Flow Model

Figure 4.21 shows a simple model of saturated funneled flow and subsequent fingering flows with a unit thickness, where the width of the controlled area l_i is assumed to be equal to the distance of fingers L_i. When a fine layer is in contact with an underlaid coarse layer and the interface is inclined, the constant vertical downward flux will accumulate above the interface and will induce saturated funneled flow along the interface. Since the cross section of the saturated funneled flow is inclined as shown in Figure 4.22, the rate of saturated funneled flow Q (m^3 s^{-1}) is given by

$$Q = -K_s\left(\frac{\partial \psi_m}{\partial X} - \sin\phi\right)\tau\cos\phi \qquad (4.14)$$

where K_s is the saturated hydraulic conductivity of funneled flow, ψ_m is the matric head, X is the distance along the inclined interface, ϕ is the inclination of the slope, and τ is the thickness of saturated funneled flow.

Since both τ and Q increase with the dip along the interface, the first fingering flow will appear at a distance of L_1 (m), where the thickness τ and rate Q will have maximum values of τ_{max} and Q_{max}, respectively. The simple model of saturated

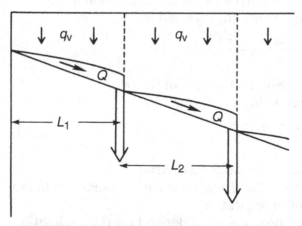

Figure 4.21 Simple model of saturated funneled flow and subsequent fingering flow with a unit thickness.

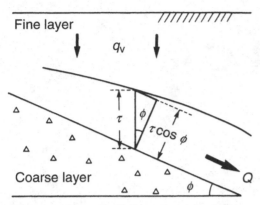

Figure 4.22 Scaled-up funneled flow.

funneled flow assumes that the mass balance equation for fingering flow is given by

$$Q_{max} = q_v L_1 \tag{4.15}$$

where q_v is the constant vertical flux (m s^{-1}). Hence, the distance L_1 is equal to the value Q_{max}/q_v. Taking account of the assumption that funneled flow is saturated in this model, the matric head gradient $\partial \psi_m / \partial X$ is zero. The maximum rate of funneled flow is, therefore, given by

$$\begin{aligned}
Q_{max} &= K_s \sin \phi \, \tau_{max} \cos \phi \\
&= \frac{K_s \tau_{max} \sin 2\phi}{2}
\end{aligned} \tag{4.16}$$

Figure 4.23 gives a classic capillary model of water retention in a thin tube open to a thick tube in which water is held by capillary force. The condition for holding the capillary water is given by

$$\rho_w g \tau \leq 2 \sigma_w \left(\frac{1}{r} - \frac{1}{R} \right) \tag{4.17}$$

where ρ_w is the density of water, g is the constant of gravity, τ is the height of capillary water (which is assumed to be equal to the thickness of funneled flow), σ_w is the surface tension of water, r is the radius of the thin tube, and R is the radius of the thick tube. By applying this capillary water model to

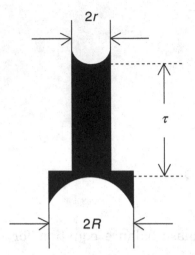

Figure 4.23 Capillary model of water retention in a thin tube open to a thick tube.

saturated funneled flow at the interface, Kung (3) defined the maximum thickness τ_{max} of saturated funneled flow by

$$\tau_{max} = \frac{2\sigma_w}{\rho_w g}\left(\frac{1}{r} - \frac{1}{R}\right) \tag{4.18}$$

where r is the equivalent radius of the finer capillary and R is the equivalent radius of the coarser capillary. Substituting Equation (4.18) into Equation (4.16), we obtained the maximum rate of funneled flow as

$$Q_{max} = \frac{K_s \sigma_w \sin 2\phi}{\rho_w g}\left(\frac{1}{r} - \frac{1}{R}\right) \tag{4.19}$$

Substituting Equation (4.19) into Equation (4.15), we eventually obtained the distance where the first fingering will appear as

$$L_1 = \frac{K_s \sigma_w \sin 2\phi}{q_v \rho_w g}\left(\frac{1}{r} - \frac{1}{R}\right) \tag{4.20}$$

The thickness of funneled flow beyond the distance L_1 may be zero but will increase again with distance downward along the interface until the thickness of τ reaches τ_{max}, when the second finger will occur. Funneled flow and subsequent fingering flows will thus repeat at regular intervals unless boundary conditions affect the flow.

2. Unsaturated Funneled Flow Model

Ross (19) investigated funneled flow theoretically, referring to subsequent fingering flow as a capillary barrier because the controlling physical mechanism is the capillary pressure developed in the fine-grained soil, which prevents water from entering the larger pores of the coarse soil. Figure 4.24 denotes the unsaturated flux of water q and its components in the top fine soil layer, where q_x is the component in the direction of the x-axis that is parallel to the interface, q_z is the component in the direction of the z-axis whose origin is equal to the interface, and q_h is the horizontal component of q. Ross (19) assumed that the integrated value of q_h in the top fine soil layer is related to funneled flow and subsequent fingering flow.

The component q_h is given by

$$q_h = q_x \cos \phi + q_z \sin \phi \qquad (4.21)$$

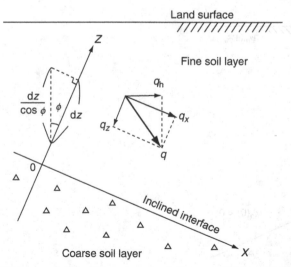

Figure 4.24 Unsaturated flux and its components.

where q_x and q_z are defined by

$$q_x = -K\left(\frac{\partial \psi_m}{\partial X} - \sin\phi\right) \qquad (4.22)$$

$$q_z = -K\left(\frac{\partial \psi_m}{\partial Z} + \cos\phi\right) \qquad (4.23)$$

in which K is the unsaturated hydraulic conductivity of fine soil, ψ_m is the matric head of water, and ϕ is the inclination of the interface. Substituting Equations (4.22) and (4.23) into Equation (4.21) and assuming that $\partial\psi_m/\partial X$ is zero in the capillary barrier (funneled flow), we obtain

$$q_h = -K\frac{\partial \psi_m}{\partial Z}\sin\phi \qquad (4.24)$$

The total amount of funneled flow is given by integrating Equation (4.24) in the vertical direction as shown in Figure 4.25 such that

$$Q = \int q_h(z)\frac{dZ}{\cos\phi} \qquad (4.25)$$

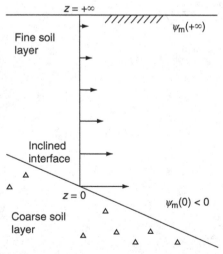

Figure 4.25 Vertical distribution of the component q_h.

where Q is the integrated flow of water in the horizontal direction. Substituting Equation (4.24) into Equation (4.25), the integration becomes

$$Q = -\tan\phi \int K \, d\psi_m \tag{4.26}$$

Ross (19) assumed that the unsaturated hydraulic conductivity is given by

$$K = K_s \exp(b\psi_m) \tag{4.27}$$

where K_s is the saturated hydraulic conductivity and b is the parameter determined experimentally. Substituting Equation (4.27) into Equation (4.26), the integrated flow is denoted by

$$Q = -K_s \tan\phi \int \exp(b\psi_m) \, d\psi_m \tag{4.28}$$

The maximum value of Q is given by integrating from ψ_m at $Z = 0$ to ψ_m at $Z = \infty$ such that

$$Q_{max} = -K_s \tan\phi \int_{\psi_m(0)}^{\psi_m(\infty)} \exp(b\psi_m) \, d\psi_m$$
$$= -\frac{\tan\phi}{b}\{K[\psi_{m(\infty)}] - K[\psi_{m(o)}]\} \tag{4.29}$$

Since $K[\psi_m(0)]$ is always less than K_s and $K[\psi_m(\infty)]$ is regarded to be equal to q_v, the vertical downward flux near the soil surface being positive regardless of the axes, Ross (19) obtained the diversion capacity

$$Q_{max} < \frac{\tan\phi}{b}(K_s - q_v) \tag{4.30}$$

Applying this diversion capacity to Equation (4.15), we can obtain an estimation of the distance L_1 where the first finger will appear in the coarse soil layer such that

$$L_1 < \frac{\tan\phi}{b}\left(\frac{K_s}{q_v} - 1\right) \tag{4.31}$$

3. Further Extension of the Models

Three major points remain to be considered when determining whether to apply the saturated model given by Equation (4.20) or the unsaturated model given by Equation (4.31) to the actual funneled flows. The first is how to determine the equivalent radius of finer capillary r and the equivalent radius of coarser capillary R in the saturated funneled flow model. One of the most reliable estimations of the equivalent radius of a capillary was given by Tabuchi (23), who showed theoretically that the menisci of water in pores of sphere-packed materials are spheroidal rather than spherical. Taking account of this theory, the value of r or R (cm) is estimated by

$$r \text{ or } R = \frac{0.12}{h_e} \qquad (4.32)$$

where h_e is the air entry suction of the sphere-packed material. For example, when h_e of a sphere-packed material is 35 cm, its equivalent radius is estimated to be 0.00343 cm; when h_e is only 1 cm, its equivalent radius is 0.12 cm. Thus, if the latter material is overlaid by the former, the value r is 0.00343 cm and R is 0.12 cm, resulting in a τ_{max} value of 42.0 cm.

Another estimation of r and R is given by using the scaling theory of the similar-media concept (see Chapter 9). When a soil denoted A is assumed to be similar to a comparable soil B whose saturated hydraulic conductivity $K_s(B)$ and air entry suction $h_e(B)$ are known, the unknown air entry suction $h_e(A)$ is estimated using the scaling theory (24)

$$K_s(A)h_e(A)^2 = K_s(B)h_e(B)^2 \qquad (4.33)$$

where $K_s(A)$ is the saturated hydraulic conductivity of soil A. For example, if $K_s(B)$ is 5×10^{-3} cm s^{-1} with a $h_e(B)$ value of 35 cm, and if $K_s(A)$ is 10^{-3} cm s^{-1}, the value of $h_e(A)$ is estimated to be 78.3 cm, and using Equation (4.32), the equivalent radius in the pores of soil A is determined to be 0.0015 cm. Many other methods of estimating the equivalent radius of r and R may be useful. It is important, however, to note that the value of L_1 in Equation (4.20) is very sensitive to the values r and R, and therefore these parameters have to be determined very carefully.

It remains to determine the parameter b in Equation (4.27) and to investigate the sensitivity of the resultant equation (4.31) to the value b in the unsaturated funneled flow model. When a soil has a distinct air entry value ψ_{me} (negative value) in its soil moisture characteristic curve, the unsaturated hydraulic conductivity should be, as pointed out by Steenhuis et al. (25), approximated by

$$K = \begin{cases} K_s & \text{for } 0 \leq h \leq h_e \\ K_s \exp[-b(h + h_e)] & \text{for } h_e < h \end{cases} \tag{4.34}$$

By using Equation (4.34) instead of Equation (4.27), Equation (4.31) is replaced by

$$L_1 < \tan \phi \left[\frac{1}{b} \left(\frac{K_s}{q_v} - 1 \right) + \frac{K_s}{q_v} (h_e - h_w) \right] \tag{4.35}$$

where h_w is the water entry suction of the coarse layer. Since the value of L_1 is very sensitive to the value b, the appropriate adoption of the unsaturated hydraulic conductivity function is crucial in the unsaturated funneled flow model.

It also remains to measure L_1 experimentally. When we measure L_1 from, say, Figure 4.17 or Figure 4.18, the difference between the diameters of the fingers and the thickness of the Hele–Shaw model must be taken into account. Figure 4.26 is a three-dimensional illustration of the funneled flow and a fingering flow, where W is the width of the Hele–Shaw model and f is the diameter of the finger. If W is equal to f, the flows may have the characteristics of two-dimensional flows in which all the funneled flow will be directed into the finger. If, on the other hand, W is larger than f, the funneled flow will be concentrated into the finger and the effect of the concentration will increase with the increased difference between W and f. The new factor, defined by

$$\xi = \frac{W}{f} \tag{4.36}$$

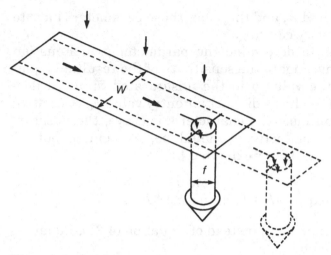

Figure 4.26 Relation between the width of funneled flow W and the width of fingering flow f.

is useful to correct the measured value L_1, and ξ is termed a catchment factor. When the value of ξ is large, the location where the fingering occurs will approach the upside of the sloping interface, whereas when the value of ξ is small, the location where the fingering occurs will be down away along the sloping interface. When the value of ξ is below 1, the diameter of the finger is restricted by the width of the Hele–Shaw model and this type of correction may not be applicable. Thus, experimental values of L_1 are corrected by multiplying the catchment factor ξ for a comparison of saturated and unsaturated funneled flow models, both of which are two-dimensional models.

E. Relation between Theory and Application

Both the saturated funneled flow model and the unsaturated funneled flow model provide the theoretical prediction of the horizontal distance L_1, the distance between the left sidewall and the location where the first finger appears. All the experimental values obtained from Runs 1 to 9 are comparable with these model calculations.

The parameter values in Equations (4.20) and (4.35) were determined experimentally. Table 4.4 gives these values for Runs 1 through 9 in Table 4.3. The extended modifications of the funneled flow models given above were used to determine the parameters. The values of L_1 were obtained by multiplying the catchment factor ξ by each measured distance of the first fingers in each Hele–Shaw model. Figure 4.27 shows the comparison between predicted and measure distances L_1 of

Table 4.4 Values of Parameters in Equations (4.20) and (4.35)

Parameters		Value
K_s	Saturated hydraulic conductivity of sand	2.45×10^{-2} cm s^{-1}
σ_w	Surface tension of water at 20°C	$72.75 \ N \ m^{-1}$
ρ_w	Density of water	1 g cm^{-3}
g	Constant of gravity	980 cm s^{-2}
R	Equivalent radius of coarser capillary	0.0239 cm
r	Equivalent radius of finer capillary	0.00467 cm
b	Experimental constant in Equation (4.29)	0.245
h_e	Air entry suction of finer layer	27.4 cm
h_w	Water entry suction of coarser layer	5 cm

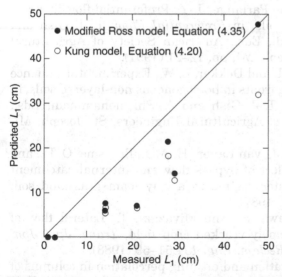

Figure 4.27 Comparison between predicted and measured distance L_1 of capillary barrier. (After Miyazaki, T., *Trans. Jpn. Soc. Irrig. Drain. Reclam. Eng. 179*:49–56 (1995). With permission.)

each capillary barrier. Even though the modified Ross model (Equation (4.35)) gives slightly better predictions to the measured values than Kung model (Equation (4.20)) does, both predictions are obviously underestimating the measured values. These underestimations may be attributed to the neglect of curving of vertical down streamlines in these models that appear in the upper layer, typically seen in Figure 4.17 and Figure 4.18.

REFERENCES

1. Beven, K., Modeling preferential flow: an uncertain future? in *Preferential Flow,* T. J. Gish and A. Shirmohammadi, Eds., American Society of Agricultural Engineers, St. Joseph, MI, pp. 1–11 (1991).
2. Kung, K.-J. S., Preferential flow in a sandy vadose zone. 1. Field observation, *Geoderma 46*:51–58 (1990).
3. Kung, K.-J. S., Preferential flow in a sandy vadose zone. 2. Mechanism and implications, *Geoderma 46*:59–71 (1990).
4. Beven, K. and Germann, P., Macropores and water flow in soils, *Water Resour. Res. 18*(5):1311–1325 (1982).
5. Steenhuis, T. S. and Parlange, J.-Y., Preferential flow in structured and sandy soils, in *Preferential Flow,* T. J. Gish and A. Shirmohammadi, Eds., American Society of Agricultural Engineers, St. Joseph, MI, pp. 12–21 (1991).
6. Hendrickx, J. M. H. and Dekker, L. W., Experimental evidence of unstable wetting fronts in homogeneous non-layered soils, in *Preferential Flow,* T. J. Gish and A. Shirmohammadi, Eds., American Society of Agricultural Engineers, St. Joseph, MI, pp. 22–32 (1991).
7. van Stiphout, T. P. J., van Lanen, H. A. J., Boersma, O. H., and Bouma, J., The effect of bypass flow and internal catchment of rain on the water regime in a clay loam grassland soil, *J. Hydrol. 95*:1–11 (1987).
8. Inoue, H., Hasegawa, S., and Miyazaki, T., Lateral flow of water in an extremely cracked crop field, *Trans. Jpn. Jpn. Soc. Irrig. Drain. Reclam. Eng. 134*:51–59 (1988).
9. Tabuchi, T., Infiltration and ensuing percolation in columns of layered glass particles packed in laboratory, *Trans. Agric. Eng. Soc. Jpn. 1*:13–19 (1961).

10. Hill, D. E. and Parlange, J.-Y., Wetting front instability in layered soils, *Soil Sci. Soc. Am. Proc. 36*:697–702 (1972).
11. Raats, P. A. C., Unstable wetting fronts in uniform and nonuniform soils, *Soil Sci. Soc. Am. Proc. 37*:681–685 (1973).
12. Philip, J.R., The growth of disturbances in unstable infiltration flows, *Soil Sci. Soc. Am. Proc. 39*:1049–1053 (1975).
13. Diment, G. A. and Watson, K. K., Stability analysis of water movement in unsaturated porous materials. 3. Experimental studies, *Water Resour. Res. 21*:979–984 (1985).
14. Tamai, N., Asaeda, T., and Jeevaraj, C. G., Fingering in two-dimensional, homogeneous, unsaturated porous media, *Soil Sci. 144*(2):107–112 (1987).
15. Baker, R. S. and Hillel, D., Laboratory tests of a theory of fingering during infiltration into layered soils, *Soil Sci. Soc. Am. Proc. 54*:20–30 (1990).
16. Kawamoto, K. and Miyazaki, T., Fingering flow in homogeneous sandy soils under continuous rainfall infiltration, *Soils Found. 39*(4):79–91 (1999).
17. Gardner, W. H., *Water Movement in Soil,* film, Washington State University, Pullman, WA (1960).
18. Miyazaki, T., Water flow in unsaturated soil in layered slopes, *J. Hydrol. 102*:201–214 (1988).
19. Ross, B., The diversion capacity of capillary barriers, *Water Resour. Res. 26*(10):2625–2629 (1990).
20. Miyazaki, T., Refraction, fingering and lateral flow of water in layered soils, *EOS Trans. Am. Geophys. Union 71*(28):869 (1990).
21. Miyazaki, T., Diversion capacity of capillary barrier in a layered slope, in *Trends in Hydrology,* S. G. Pandalai, Ed., Council of Scientific Information, Trivandrum, pp. 469–475 (1994).
22. Miyazaki, T., The diversion capacity of an inclined capillary barrier, *Trans. Jpn. Soc. Irrig. Drain. Reclam. Eng. 179*:49–56 (1995).
23. Tabuchi, T., Theory of suction drain from the saturated ideal soil, *Soil Sci. 102*:161–166 (1966).
24. Campbell, G. S., *Soil Physics with Basic,* Elsevier, New York, p.52 (1985).
25. Steenhuis, T. S., Parlange, J.-Y., and Kung, K.-J. S., Comment on "The diversion capacity of capillary barriers" by Benjamin Ross, *Water Resour. Res. 27*(8):2155–2156 (1991).

5

Water Flow in Slopes

I.   WATER BALANCE ON SLOPES

Water balance is defined as the quantitative relation among integrated input (Q_{in}), output (Q_{out}), and storage (ΔS) in a given space and a given time, and is described by

$$Q_{in} = Q_{out} + \Delta S \tag{5.1}$$

The required space for defining a water balance is generally called the *representative element volume* (REV). When the REV is given along a slope, as shown in Figure 5.1, the water balance is

$$P + I_r + R_{in} + L_{in} = (E + T + R_{out} + L_{out} + Q_d)$$
$$+ (\Delta S_1 + \Delta S_2) \tag{5.2}$$

where the input is composed of the precipitation, P, the quantity of irrigation, I_r, the inflow of surface water, R_{in}, and the lateral inflow of soil water, L_{in}; the output is composed of evaporation, E, transpiration, T, surface runoff, R_{out}, lateral outflow of soil water, L_{out}, and downward drainage, Q_d. The change of storage is composed of two terms, surface storage, ΔS_1, and subsurface storage, ΔS_2.

When the REV is divided by the land surface, the balance of surface water and the balance of soil water are given separately. The balance of surface water is

$$P + I_r + R_{in} = E + T + R_{out} + I + \Delta S_1 \tag{5.3}$$

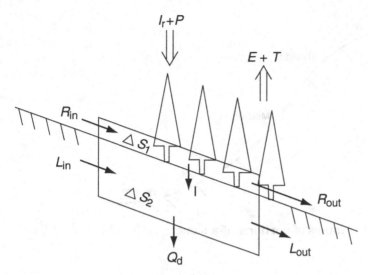

Figure 5.1 Water balance of a representative element volume along a slope.

and the balance of soil water is

$$I + L_{\mathrm{in}} = L_{\mathrm{out}} + Q_{\mathrm{d}} + \Delta S_2 \tag{5.4}$$

where I denotes infiltration from the surface into slopes. When the water balance is integrated over a long term, say 1 year, any changes in ΔS_1 and ΔS_2 are eliminated.

II. MOISTURE CONDITIONS IN SLOPES

A. Topographical Effects on Moisture Condition

Typical slopes consist of top, shoulder, midslope, foot, and bottom, as shown in Figure 5.2, and are classified according to their degree of incline in relation to flat land (less than about 3°): gentle slope, slope, steep slope, very steep slope, and precipice (greater than 60°). Generally speaking, the top and shoulder of a slope are drier and the foot and bottom of a slope are moister.

Figure 5.3 shows an example of the distribution of moisture with the bulk density of surface soil along a natural midslope of 25° covered with grass (1). Along the slope no

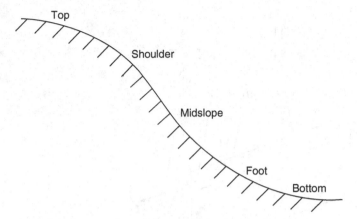

Figure 5.2 Names of portions of a slope.

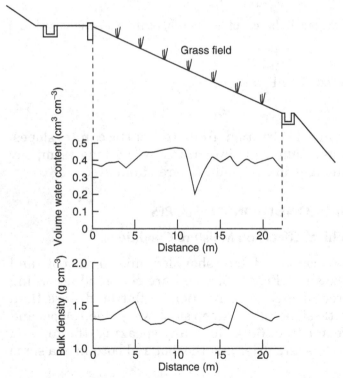

Figure 5.3 Moisture condition of surface soil along a natural midslope. (After Miyazaki, T., *Pedologist* 31(2):54–64 (1987). With permission.)

tendencies are evident for either volumetric water content or bulk density. However, even a small topographical alteration through cutting or banking of soil influences the moisture condition of the slope.

Figure 5.4 shows suction changes in surface soils at the cut side (solid line) and the banked side (broken line) (2). When soil was continuously in a drying state, suction on the banked side was higher than suction on the cut side, where as when the soil was wet by a rain on July 30, suction on the banked side decreased less than did suction on the cut side. The change in soil structure caused by artificial banking may have caused the easy-to-dry soil and easy-to-wet soil in the slope.

The existence of an impermeable layer clearly dominates the flow of water in a slope. Ooeda et al. (3) poured ^{60}Co onto a slope having an impermeable layer at a depth of 30 cm, as shown in Figure 5.5, and traced the lateral water flow by monitoring the radiation counts of soil samples. The numbered curves in Figure 5.5 are contours of the activities

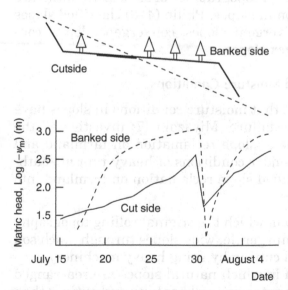

Figure 5.4 Moisture condition of surface soil along a cut and banked slope. (After Matsuda, M. and Yamada, N., *Bull. Kagawa Univ. Fac. Agric.* 22(2):113–117 (1971). With permission.)

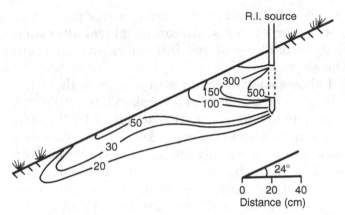

Figure 5.5 Contours of the activities (cpm) of ^{60}Co in a slope. (After Ooeda, M., Fujioka, Y., Katsurayama, K., and Tajima, S., *Trans. Agric. Eng. Soc. Jpn. 2*:75–81 (1961). With permission.)

(counts per minute [cpm] per 20 g of dry soil). It was confirmed that subsurface lateral flow was predominant just above the impermeable layer in this slope.

Concave and convex topographies of slopes also influence the moisture condition in slopes. Philip (4–6) classified slopes as planner slopes, divergent slopes, convergent slopes, concave slopes, and convex slopes.

B. Slope Failures and Moisture Conditions

It is often pointed out that moisture conditions in slopes have an influence on slope failure. Miyazaki (7) investigated the relation between type of slope reclamation on farmland and type of slope failure under conditions of heavy rain in southwest Japan. He classified slope reclamation on farmland into four types:

1.   Reclamation in which the original rolling topography is changed into gentle, wide slopes through much soil banking and cutting by using heavy machinery.
2.   Reclamation in which natural slopes are rearranged through less extensive soil banking and cutting than that in type 1, to construct operating roads and operating facilities.

3. Reclamation in which natural slopes are used to the extent possible and the slopes undergo only partial amendment for the construction of operating roads and facilities.
4. Terracing reclamation in which slopes are developed in small flat fields through small amounts of soil banking and cutting.

The slope reclamation methods listed above are those typically used in Asian countries. Miyazaki classified slope failure into the following three categories (Figure 5.6):

1. Large-scale slope failure
2. Collapse at the foot of long slopes
3. Slide or collapse of small slopes

Table 5.1 is a matrix of the number of slope failures classified by type of slope reclamations and slope failure. It is evident that a correlation exists: the larger the scale of reclamation, the larger the scale of slope failure. In elaborate

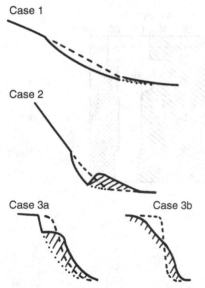

Figure 5.6 Three categories of slope failures in Mitoyo County, Japan. (After Miyazaki, T., *Jpn. Agric. Res. Q. 20*(3):174–179 (1987). With permission.)

Table 5.1 Classification of Slope Failure by Type of Slope Reclamation and Type of Slope Failure

Type of reclamation	Type of slope failure		
	1	2	3
1	25	0	0
2	19	15	0
3	8	14	12
4	0	3	0
Total		96	

field investigations of large-scale slope failures, plant layers sandwiched between the original slopes and the banked soils were observed in all cases. These layers had presumably been buried during large-scale reclamation of the slopes (type 1).

Figure 5.7 shows the average volume fractions of soils in the vicinity of the sandwiched plant layers, whose average

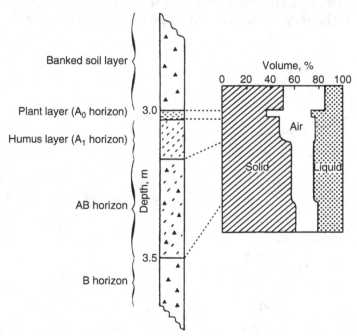

Figure 5.7 Average volume fractions of soils in the vicinity of sandwiched plant layers. (After Miyazaki, T., *Jpn. Agric. Res. Q.* *20*(3):174–179 (1987). With permission.)

thickness is about 3 cm. The plant layers have small solid-phase fractions, high porosity values, and high saturated hydraulic conductivity values. Miyazaki (7) suggested that a significant relation may exist between large-scale slope failure and the existence of buried plant layers in large-scale-reclaimed slopes.

Figure 5.8 shows experimental results for slope failure in a model slope of sandy loam, in which a plant layer 3-cm thick is buried artificially. The model slope, whose inclination is about 15°, is 1.6-m wide, 1-m deep, and 0.2-m thick. Water was supplied by a perforated polyvinyl pipe inserted vertically in the upside of the slope, as shown in the figure. The entire slope was saturated with water 24 h before the experiment and was subsequently drained from the bottom. It was observed in this experiment that water concentrated in the sandwiched layer, washing away the sandy loam around the plant layer, resulting finally in slope failure at the lower tip of the model slope, as shown in Figure 5.8, where the initial form of the slope is denoted by dashed lines. The effects of such a

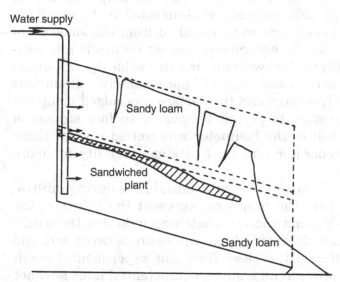

Figure 5.8 Concentration of lateral water flow in a sandwiched plant layer and the resultant digging of soil in a model slope whose initial form is denoted by the dotted line.

sandwiched plant layer on the stability of natural slopes are, however, still somewhat vague.

III. INFILTRATION INTO SLOPES

A.   Potential Profiles During Infiltration into Slopes

In a rainstorm, when infiltration begins on a slope, the wetting front, and therefore the contour of matric potential, advance downward, remaining parallel to the surface of the slope. Figure 5.9 shows a wetting front 200 min after the start of infiltration in a model slope of sandy loam under an artificial rainfall of 20 mm h^{-1}. The model slope is the same as that used in Figure 5.8. The wetting front retained its shape parallel to the slope.

To determine more clearly the shape of the wetting front in a slope, Miyazaki (1) measured changes in the two-dimensional matric head profile during infiltration under an artificial rainfall of 20 mm h^{-1} by using a slope-adjustable lysimeter 10-m long, 2-m wide, and 1-m deep and a rain generator with 2500 nozzles, as illustrated in Figure 5.10, where 100 tensiometers were installed from the sidewall of the lysimeter. All the equipment was set up inside a laboratory. Andisol (light clay) with an air entry value ψ_{me} of -40 cm and a water entry value ψ_{mw} of approximately -20 cm was packed in the lysimeter and the surface was ridged along the contour of the slope to permit temporary surface storage of rainfall. The soil in the lysimeter was wetted several times by the rain generator and kept in static stability for more than 1 month.

Figure 5.11 shows the equipotential lines during infiltration by rain; here the numbers represent the values of the matric heads ψ_m and the crosshatching indicates the saturated zone, where the matric head values are between zero and ψ_{mw} (i.e., -20 cm). It is clear from this experimental result that in a uniform soil on a slope, equipotential lines advance downward, maintaining the shape parallel to the slope during infiltration by rain. This feature may be applicable to the top,

Figure 5.9 Wetting front, designated by the interface between the dark zone (wet zone) and the light zone (dry zone), after 200 min of the start of infiltration in a model slope of sandy loam.

middle, and bottom parts of natural slopes but may not be applicable to the shoulder or foot of natural slopes, where the concentration or diversion of subsurface water flow has a greater influence on the advancement of equipotential lines during infiltration.

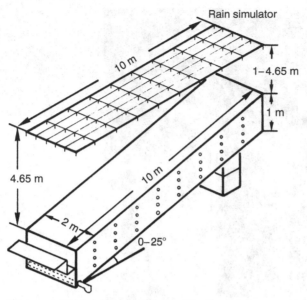

Figure 5.10 Slope-adjustable lysimeter equipped with artificial rain generator.

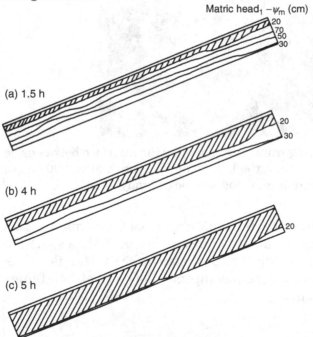

Figure 5.11 Movement of equipotential lines during infiltration under an artificial rainfall of $20\,\mathrm{mm\,h^{-1}}$.

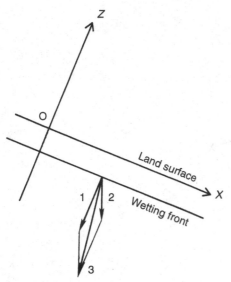

1 Gradient of matric head
2 Gravity
3 Resultant force

Figure 5.12   Driving force of infiltration and components.

B.   Flux During Infiltration into Slopes

The driving force of infiltration is composed of two components, the gradient of matric potential and the gravity, as shown in Figure 5.12, where the driving force is shown between the two forces. Figure 5.13 shows the conceptual relation between the flux of rain **r** and the flux of infiltrated water **q** in the slope, where the x-axis is positive along the downslope of the surface and the z-axis is positive upward rectangular to the slope surface. It is assumed that the flux of rain is less than the infiltration capacity of the soil, which is defined as the flux across the soil surface under ponded water. If the flux of rain is greater than the infiltration capacity, the excess water will cause surface runoff on the slope.

In Figure 5.13, the continuity of the flux of rain **r** and the flux of infiltration **q**$(z = 0)$ at the surface of the slope is described by

(a)

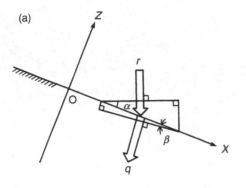

(b)

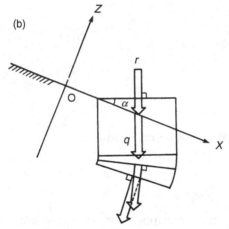

Figure 5.13 Refraction of flux of water in (a) early stage and (b) subsequent stage of infiltration.

$$r \cos \alpha = q(z = 0) \cos \beta \tag{5.5}$$

where r and q are the absolute values of fluxes $\mathbf{r}$ and $\mathbf{q}$, respectively, α is equal to the angle of the slope ϕ, and β is the refraction angle at the soil surface, the latter being equal to the angle between the soil surface and a section normal to the flux $\mathbf{q}(z = 0)$. Assuming that the soil is isotropic, the flux $\mathbf{q}$ at any depth in the slope is decomposed into two components,

$$q_x = -K\left(\frac{\partial \psi_m}{\partial x} - \sin \alpha\right) \tag{5.6}$$

$$q_z = -K\left(\frac{\partial\psi_m}{\partial z} + \cos\alpha\right) \tag{5.7}$$

where q_x is the component in the direction of the x-axis, q_z is the component in the direction of the z-axis, and K is the unsaturated hydraulic conductivity, which is a function of the matric head ψ_m of the soil water. It is evident that the flux of water at the soil surface of the slope $\mathbf{q}(z = 0)$ is also decomposed into Equations (5.6) and (5.7), and their ratio is given by

$$\frac{q_x}{q_z} = \tan\beta \tag{5.8}$$

Substitution of Equations (5.6) and (5.7) into Equation (5.8) yields

$$\tan\beta = -\frac{\sin\alpha}{\cos\alpha + (\partial\psi_m/\partial z)_{z=0}} \tag{5.9}$$

This equation means that at the beginning of a rainstorm, when the matric head gradient $(\partial\psi_m/\partial z)_{z=0}$ is large enough at the surface of the slope, a refraction of flux occurs at the soil surface as shown in Figure 5.13(a).

When the wetting front advances deeply, the matric head gradient at the soil surface $(\partial\psi_m/\partial z)_{z=0}$ may be very small and the incidence angle α of rain and the refraction angle β of the flux in the soil may be almost equal, as shown in Figure 5.13(b). Since the wetting front advances downward, maintaining the front parallel to the slope surface during infiltration, the matric head gradient in the vicinity of the wetting front is directed inside the slope and the flux there may bend inside the slope as shown in Figure 5.13(b). The directions of all the fluxes in the slope are given by

$$\gamma = \tan^{-1}\frac{q_x}{q_z} \tag{5.10}$$

where γ is the angle between the z-axis and the flux q and is equal to β at the soil surface.

C. Equation of Flow

The equation of two-dimensional flow in a slope is given by

$$\frac{\partial \theta}{\partial t} = -\left(\frac{\partial q_x}{\partial x} + \frac{\partial q_z}{\partial z}\right) \tag{5.11}$$

where q_x and q_z are defined by Equations (5.6) and (5.7). In a long slope, the matric head gradient along the x-axis, $\partial \psi_m/\partial x$, may be zero during infiltration since the wetting front is parallel with the slope surface during infiltration. Hence, substituting Equations (5.6) and (5.7) into Equation (5.11), we obtain the one-dimensional equation

$$\frac{\partial \theta}{\partial t} = \frac{\partial}{\partial z}\left(K\frac{\partial \psi_m}{\partial z}\right) + \cos\alpha\left(\frac{\partial K}{\partial z}\right) \tag{5.12}$$

The boundary condition of a constant rain intensity r and the initial condition of a constant volumetric water content θ_0 are given by

$$r\cos\alpha = q_z, \quad z = 0,\, t > 0 \tag{5.13a}$$

$$\theta = \theta_0, \quad z \le 0,\, t = 0 \tag{5.13b}$$

By substituting Equation (5.7) into Equation (5.13a), the boundary condition can be rewritten as

$$\left(\frac{\partial \psi_m}{\partial z}\right)_{z=0} = -\cos\alpha\left(\frac{r}{K} + 1\right) \tag{5.14}$$

Assuming that the slope is long, homogeneous, and semi-infinite, Equation (5.12) subjected to the initial condition (5.13b) and the boundary condition (5.14) is solved approximately by a numerical method.

Figure 5.14 shows an example of numerical solutions for various angles of a slope of sandy loam. A simple explicit finite-difference technique was used for solving Equation (5.12). The input data in the calculation were given from experimental measurements. The rain intensity r was $20\,\text{mm}\,\text{h}^{-1}$, the initial volumetric water content θ_0 was $0.01\,\text{cm}^3\,\text{cm}^{-3}$, and the saturated volumetric water content θ_s

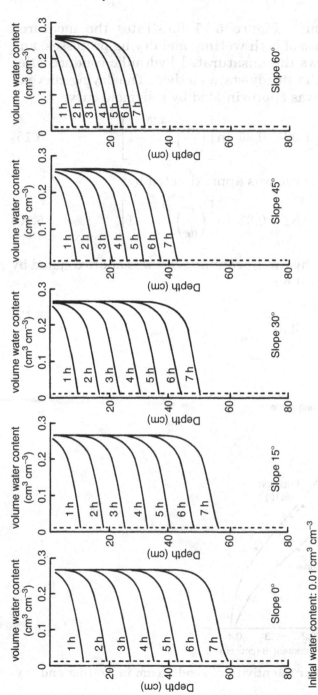

Initial water content: 0.01 cm³ cm⁻³

Rain intensity: 20 mm h⁻¹

Figure 5.14 Numerical solutions of Equation (5.12) for the various angles of a slope.

was $0.414\,\mathrm{cm}^3\,\mathrm{cm}^{-3}$. Figure 5.15 illustrates the measured water retentivities of both wetting and drying processes, and Figure 5.16 shows the unsaturated hydraulic conductivities measured with the steady-state method. The drying curve of water retention was approximated by a sigmoid curve,

$$\log(-\psi_m) = 1.00 + 0.35\,\ln\left[\left(\frac{\theta}{\theta_s}\right)^{-3.10} - 1\right] \tag{5.15}$$

and the wetting curve was approximated by

$$\log(-\psi_m) = 0.82 + 0.35\,\ln\left[\left(\frac{\theta}{\theta_s}\right)^{-1.82} - 1\right] \tag{5.16}$$

The unsaturated hydraulic conductivity was approximated by the empirical equation

$$K = \frac{12.5}{(-\psi_m)^4 + 4460} \tag{5.17}$$

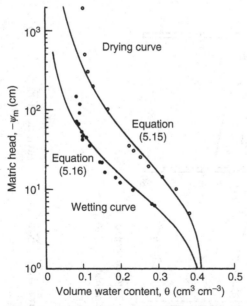

Figure 5.15 Water retentivity of sandy loam in wetting and drying processes.

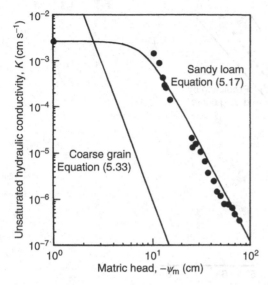

Figure 5.16 Unsaturated hydraulic conductivity of sandy loam and coarse grain.

A simulation was conducted using both the drying curve (5.15) and the wetting curve (5.16) because it has not necessarily been confirmed which retention curve is applicable for infiltration. One simple comparison of their applicability is the investigation of experimental and simulated downward velocities of wetting-front advancement. Figure 5.17 shows the depth of wetting front versus the duration of infiltration in a 15° slope, where the experimental wetting front was defined as the visual boundary between wetted and dry layers in Figure 5.9 and the simulated wetting fronts were defined as the intersection of the volumetric water content profile and the profile of initial water content. It is evident that the experimental velocity of the wetting front lies between their simulated velocities, obtained using the drying and wetting retention curves, and that the former gives the more desirable estimation. Hence, the simulated profiles in Figure 5.14 were calculated using a sigmoid approximation for the drying retention curve given by Equation (5.15). Water retentivity during downward infiltration may be rather higher than

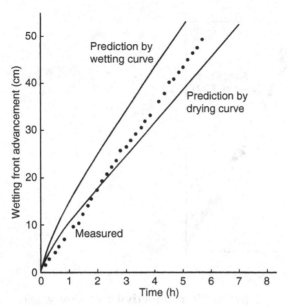

Figure 5.17 Simulated and measured velocity of wetting front advancement into a 15° slope.

that during upward wetting, which is generally used to obtain wetting retention curves. Practically, the use of drying water retention curves is recommended when one is dealing with downward infiltration.

It should be noted in Figure 5.14 that there are no significant differences in water profile advancement among slopes of less than 30°, but there are quite remarkable differences in them among slopes steeper than 30°. This evidence corresponds closely to the suggestions by Redinger et al. (8), who concluded, through numerical solution of the equation of vertical infiltration, that the effects of slopes of less than 30° are negligible for vertical infiltration. Through theoretical solution of the infiltration equation, Philip (4) also concluded that there is no significant difference among infiltrations into slopes of less than 30°.

D. Infiltration into Layered Slopes

1. Behavior of Wetting Front in a Layered Slope

Natural slopes are stratified to a greater or less degree due to organic movement, crustal movement, sedimentation, weathering, erosion, and decomposition of organic matters throughout their geologic history. Artificial slopes are also stratified, due to banking and cutting of natural slopes and to rolled compaction by machine traffic during the construction of artificial slopes. One typical type of stratification in artificial slopes is the sandwiched soil layer shown in Figure 5.7, where a plant layer is buried under banked soil.

Stratifications in slopes have an influence on the infiltration of water. Miyazaki (9) investigated the behavior of a wetting front in a model slope of 15°, in which a 3-cm-thick plant layer was sandwiched between layers of Masa sandy loam, under artificial rain conditions of $20\,\text{mm}\,\text{h}^{-1}$. The model slope is the same as shown by Figure 5.9. Figure 5.18 shows traces of the wetting front at 10-min intervals. Downward advancement of the wetting front retained a constant velocity during the initial 300 min. The wetting front halted when it reached the interface between the top sandy loam and the second plant layer. Most water leakage, into the second and third layers, occurred at the lower edge of the slope; slight leaks appeared at the lower part of the interface.

2. Cessation of Wetting Front at Inclined Interface

Reasons for the apparent cessation of wetting fronts at the boundary between top fine material and second-layer coarse material were given by Miller and Gardner (10), who found that the wetting front halts for a time at the horizontal interface between the top clay layer and the second, sand layer. Hillel (11) summarized their statement as follows: "Water at the wetting front is normally under suction, and this suction may be too large to permit entry into the relatively large pores of the coarse layer." The classic capillary model, illustrated in Figure 4.19 and corresponding equation (4.11), is certainly applicable to the inclined interface between the top fine

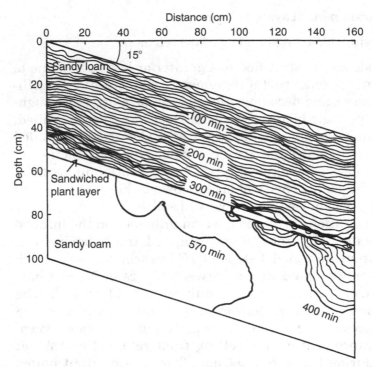

Figure 5.18 Traces of wetting front every 10 min in a model slope.
(After Miyazaki, T., *Trans. Jpn. Soc. Irrig. Drain. Reclam. Eng.*
133:1–9 (1988). With permission.)

material and the second-layer coarse material as well as to the
horizontal interface. It should, however, be noted that water
does not move during cessation of wetting-front advancement
at a horizontal interface (Figure 5.19a) and lateral water flow
takes place at an inclined interface as illustrated in Figure
5.19(b).

The mass balance of lateral water flow along an inclined
interface is denoted by the model shown in Figure 5.20, where
the x-axis is identical with the interface between the top,
finger soil and the second, coarser soil. This lateral flow was
termed funneled flow in Chapter 4. Applying the saturated
funneled flow model to this lateral flow, the hydraulic property of the flow is given as follows.

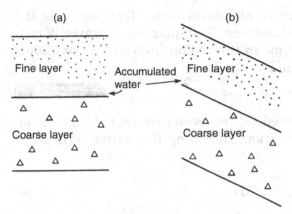

Figure 5.19 Cessation of wetting front at (a) a horizontal interface and (b) a sloping interface.

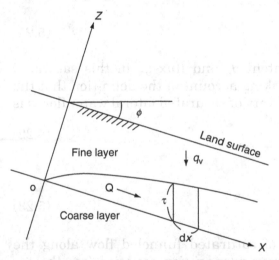

Figure 5.20 Mass balance of saturated lateral water flow along an inclined interface.

During cessation of wetting-front advancement, the top, fine soil layer is divided into two parts, a region where a constant unsaturated downward flux q_v exists and a region where a lateral saturated water flux q_x exists. The total amount of lateral water flow Q (m^3 s^{-1}) is given by

$$Q = q_x \tau \cos \phi \tag{5.18}$$

where τ is the thickness of lateral water flow and ϕ is the angle of the inclined interface. The quantity of water W contained in a small volume in the region from x to $x + dx$ shown in Figure 5.20 is given by

$$W = \theta_s \tau \, dx \, \cos \phi \qquad (5.19)$$

where θ_s is the saturated volume water content of this region. The mass balance equation, including the source term q_v, is then denoted by

$$\frac{\partial W}{\partial t} + \frac{\partial Q}{\partial x} \, dx = q_v \cos \phi \, dx \qquad (5.20)$$

Substitution of Equations (5.18) and (5.19) into Equation (5.20) yields

$$\theta_s \frac{\partial \tau}{\partial t} + q_x \frac{\partial \tau}{\partial x} = q_v \qquad (5.21)$$

where both water content θ_s and flux q_x in this saturated region are constant. Taking account of the definition that the average pore water velocity of saturated lateral water flow v is

$$v = \frac{q_x}{\theta_s} \qquad (5.22)$$

we obtain

$$\frac{\partial \tau}{\partial t} + v \frac{\partial \tau}{\partial x} = \frac{q_v}{\theta_s} \qquad (5.23)$$

This is the equation of saturated funneled flow along the inclined interface, a wave propagation equation (see the appendix of Chapter 2) whose propagation velocity is v, and is a kinematic wave equation as well as an equation of surface runoff on a slope.

As mentioned in Chapter 4, the saturated funneled flow of water along the inclined interface between the top (finer) and the second (coarser) soil layers does not continue infinitely, but leaks into the second lower layer when the flow is disturbed by an obstacle or when the thickness of the lateral flow τ is so large that condition (4.11) does not hold. This leaking flow is the fingering flow subsequent to funneled flow discussed in Chapter 4.

IV. CAUSES OF SURFACE RUNOFF

A. Infiltration Capacity and Horton Overland Flow on a Slope

Infiltration capacity was defined by Horton (12) as the maximum infiltration rate of falling rain (or melting snow) in a given soil. Since the infiltration rate is the flux of water across a land surface into soil, the maximum flux of water across a land surface into soil is equal to the infiltration capacity in flat land. However, the infiltration capacity in a slope f_t is not necessarily equal to the infiltration capacity in a flat land f_c, but generally exceeds it, especially in the early stage of infiltration.

Figure 5.21 designates the relation between infiltration capacity f_t and the maximum infiltration rate q_{max} in flat land, a gentle slope, and a steep slope. Since the flux of water in the early stage of infiltration is induced mainly by matric head gradients, the contribution of gravity is assumed to be negligible. On flat land, f_c is equal to q_{max} according to the definition of infiltration capacity. On a gentle slope, the same influx refracts at the land surface due to the refraction law (see Chapter 3), and the refracted flux is slightly less than q_{max}. On a steep slope, the same influx refracts more at the land surface, resulting in less flux than q_{max} in the soil. Note that all the crosshatched areas separated by land surfaces in Figure 5.21 (rectangle or parallelogram) are equal.

The quantitative relation between f_t and the flux $q(z = 0)$ at the soil surface is given from Equation (5.5) as

$$f_t = \frac{\cos \beta}{\cos \alpha} q(z = 0) \tag{5.24}$$

where α is the incidence angle of rain flux and β is the refraction angle of soil water flux at the surface.

In the early stage of infiltration on a slope, f_t is larger than $q(z = 0)$ because α is larger than β, as described in

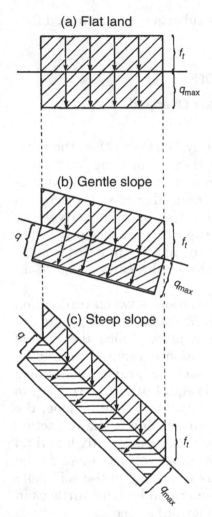

Figure 5.21 Infiltration capacity f_t and maximum infiltration rate q_{max} on (a) flat land, (b) gentle slope, and (c) steep slope.

Figure 5.13(a), but after advancement of the wetting front as shown in Figure 5.13(b), f_t is equal to $q(z = 0)$ since α is equal to β at the surface of the slope. In other words, slopes can absorb falling rain better than flat lands in the early stage of infiltration, but afterward the difference is diminished. Figure 5.22 shows typical changes in the infiltration capacities of a slope and flat land with time, both of which decrease

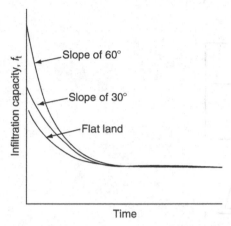

Figure 5.22 Changes in infiltration capacities of various slopes.

asymptotically to a final infiltration rate. Horton overland flow is thus defined as direct surface runoff where the rainfall intensity exceeds the maximum infiltration rate q_{max} at each slope.

When the inclination of a slope is less than about 30°, the effect of the slope on infiltration capacity is negligible, but when the slope is steeper than 30°, the slope should be considered in both infiltration capacity and Horton overland flow.

B. Return Flow

Return flow, a term of Dunne and Black (13), is the other cause of surface runoff. These authors explained stormflow hydrographs by a variable source area and overland flow concept, and attributed the unexpectedly quick increase and decrease of storm runoff in channels to the contribution of return flow. Return flow is defined more clearly as infiltrated water that returns to the land surface after having flowed a short distance into the upper soil horizon (14). Figure 5.23 shows three types of return flows that cause surface runoff. The first flows down the seepage face of groundwater on the stream bank, the second flows down the seepage face of perched water on a slope (which exists not only near the stream bank but also on midslopes), and the third is the return flow from macropore outlets in slopes. Tanaka

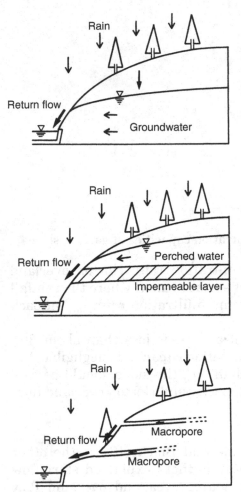

Figure 5.23   Return flows causing surface runoff.

et al. (15) showed that the majority of overland flow comprised return flow appearing at the soil surface through decayed stumps and soil pipe outlets. Figure 5.24 shows the contributing area of return flow generated both on the seepage face of groundwater and from macropore or pipe outlets. Where return flow exists, surface runoff on a slope is the sum of the return flow and rain.

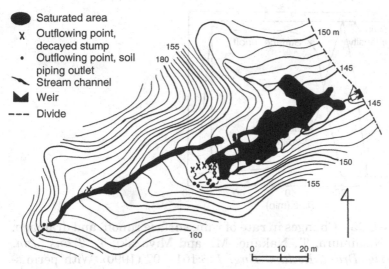

Saturated area
x Outflowing point, decayed stump
• Outflowing point, soil piping outlet
⟍ Stream channel
◼ Weir
--- Divide

Figure 5.24 Contributing area of return flow generated from outlets of macropores or pipes. (After Tanaka, T., Yasuhara, M., Sakai, H., and Marui, A., *J. Hydrol. 102*:139–164 (1988). With permission.)

C. Crust Formation

It is well known that the impact of raindrops gives rise to a crust on a soil surface. Once a surface crust has formed, the permeability of the top soil layer decreases and the infiltration capacity is lowered, resulting in surface runoff under the rainfall. Nishimura et al. (16) demonstrated the effect of crust formation on surface runoff by using a special lysimeter and a rain simulator. Andisol (clay loam), previously sieved through a 3-mm mesh screen and packed in a perforated rectangular lysimeter 30-cm wide, 50-cm long, and 10-cm deep inclined at 11°, was exposed under an artificial rain of $33\,\mathrm{mm\,h^{-1}}$. Figure 5.25 shows the artificial rain intensity, rate of surface runoff, and rate of drainage from the bottom of the lysimeter versus rainfall duration. No surface runoff and no drainage occurred in the initial 10 min while water storage in the soil increased. Ten minutes after the start of the rain, drainage from the bottom suddenly occurred, indicating that the soil was saturated and a temporary groundwater level was formed above the bottom of the lysimeter. Twenty minutes after the start of

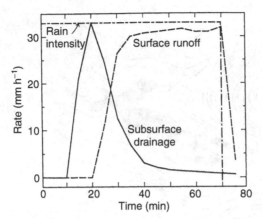

Figure 5.25 Changes in rate of rain, surface runoff, and drainage. (After Nishimura, T., Nakano, M., and Miyazaki, T., *Trans. Jpn. Soc. Irrig. Drain. Reclam. Eng. 146*:101–107 (1990). With permission.)

rain, drainage from the bottom decreased rapidly asymptotically to zero, and, simultaneously, surface runoff began, due possibly to the formation of surface crust. If the surface runoff was caused by return flow as defined above, both surface runoff and the maximum drainage from the bottom must exist, but this was not the case.

The hydraulic conductivity of the crust, shown in Figure 5.26, where the thickness of crust and subsoil are given by Δz_1

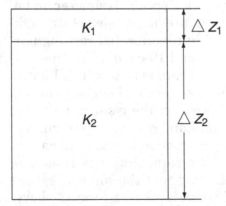

Figure 5.26 Crust and subsoil model.

and Δz_2, respectively, is estimated by the synthetic hydraulic conductivity formula of layered soil,

$$\frac{\Delta z_1 + \Delta z_2}{\langle K \rangle} = \frac{\Delta z_1}{K_1} + \frac{\Delta z_2}{K_2} \tag{5.25}$$

where $\langle K \rangle$ is the synthetic hydraulic conductivity, K_1 is the hydraulic conductivity of the crust, and K_2 is the hydraulic conductivity of the subsoil. Since the surface crust is usually very thin, it is difficult to measure K_1 directly. Instead, it is possible to estimate the value of K_1 from Equation (5.25) by measuring $\langle K \rangle$ and K_2. Figure 5.27 shows the changes in synthetic hydraulic conductivity $\langle K \rangle$ of a 5-cm-thick soil sampled from the surface of the lysimeter mentioned above during the runoff experiment, and the changes in the estimated hydraulic conductivity K_1 of a 3-mm-thick surface crust versus rainfall duration. Note that K_1 decreased with time and 50 min from the start of rainfall, the average value of K_1 decreased even less than $9.2 \times 10^{-4}\,\mathrm{cm\,s^{-1}}$ ($33\,\mathrm{mm\,h^{-1}}$). It is

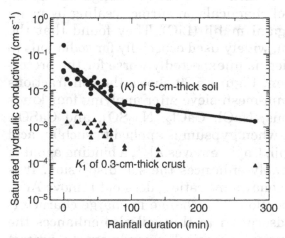

Figure 5.27 Change of synthetic hydraulic conductivity of 5-cm-thick soil and calculated hydraulic conductivity of surface crust. (After Nishimura, T., Nakano, M., and Miyazaki, T., *Trans. Jpn. Soc. Irrig. Drian. Reclam. Eng. 146*:101–107 (1990). With permission.)

clear that crust formation on a bare soil surface explains so well the occurrence of surface runoff.

Surface crust is formed mainly by being sealed with fine soil particles that have been aggregated before the rainfall at the soil surface but is dispersed by the impacts of rain drops. Nishimura et al. (17) investigated the effects of clod size, raindrop size, and initial water content on surface sealing experimentally by using a Japanese acid soil, Kunigami mahji (LiC), and a rainfall simulator. Figure 5.28 shows the relation between ratio of clods breakdown, defined by the ratio of the number of collapsed clods to the initial number of the clods, and kinetic water-drop energy of air dry, moist, and wet soils when (a) clods sizes were smaller than drop sizes, and when (b) clod sizes were larger than drop sizes. By comparing the remarkable difference between (a) and (b) of Figure 5.28, they concluded that when a soil clod is dry and smaller than a rain drop size, a slaking and resultant dispersion of fine particles occurs, while when a soil clod size is larger than a rain drop size, the clod does not easily break down and disperse.

In addition, Nishimura et al. (18) investigated experimentally the effects of chemicals on surface sealing by using the same soil, Kunigami mahji (LiC). They found that the application of gypsum, widely used especially for sodic soils to amend their properties, is unexpectedly worse for the protection against dispersion. Figure 5.29 shows the soil fractions passed through a 2-mm-mesh sieve after applying four kinds of electrolyte solutions, $NaCl$, $CaCl_2$, Na_2SO_4, and $CaSO_4$. They concluded that, when gypsum is applied to such an acid soil as Kunigami mahji, Ca^{2+} removes Al^{3+}, a binding agent of acid soils, and inevitably enhances the soil dispersion. It is notable that Na^+, a monovalent cation, does not remove Al^{3+}, resulting in less dispersion. The change in charge characteristic due to SO_4^{2-} adsorption in this soil also enhances the dispersion. In conclusion, chemical amendment to protect from crust formation on soil surface must be chosen appropriately based on the soil properties.

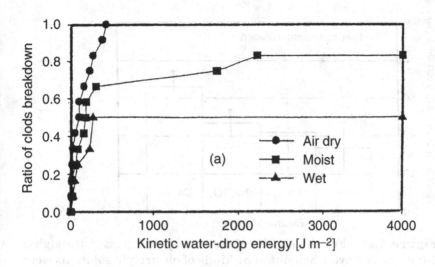

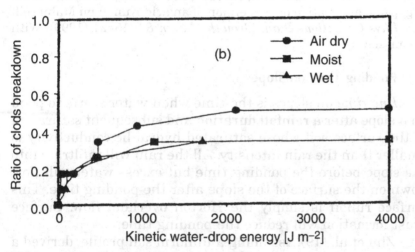

Figure 5.28 Ratio of clods breakdown versus kinetic water-drop energy for (a) smaller clod sizes than rain drop sizes, and (b) larger clod sizes than rain drop sizes. (After Nishimura, T., Nakano, M., and Miyazaki, T., Effects of clod size, raindrop size and initial moisture conditions on surface sealing of Kunigami Mahji soil, *Trans. Jpn. Soc. Irrig. Drain. Reclam. Eng.* 199:17–22 (1999). With permission.)

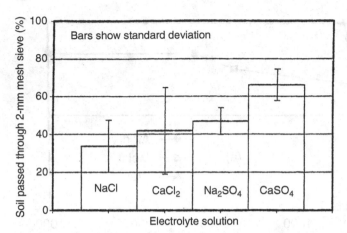

Figure 5.29 Fractions of Kunigami mahji soil passed through a 2-mm-mesh sieve when different kinds of electrolyte solutions were applied. (After Nishimura, T., Nakano, M., and Miyazaki, T., Effects of gypsum application on dispersion of an acid Kunigami Mahji soil, *Soil Phys Conditions Plant Growth, Japan 81*: 15–21 (1999). With permission.)

D. Ponding Time on Slopes

Ponding time on slopes is the time when water starts to pond on a slope after a rainfall duration and subsequent saturation of the surface soil whose saturated hydraulic conductivity is smaller than the rain intensity. All the rain will infiltrate into the slope before the ponding time but excess water will flow down on the surface of the slope after the ponding time. This surface runoff is simply the Horton overland flow. Surface crust formation will reduce the ponding time.

Zhu et al. (19), assuming a uniform soil profile, derived a modified analytical solution of ponding time t_p on a slope α, based on the original analytical solution by Kutilek (20), as

$$t_p = \frac{1}{R\cos\alpha - K(\theta_i)}$$
$$\int_{\theta_i}^{\theta_s} \frac{(\theta - \theta_i)D(\theta)}{(R\cos\alpha - K(\theta))F\Theta - (K(\theta)\cos\alpha - K(\theta_i))}\,\mathrm{d}\theta \qquad (5.26)$$

where $K(\theta)$ is the unsaturated hydraulic conductivity, $D(\theta)$ is the soil water diffusivity, R is the rain intensity, and $\Theta = (\theta - \theta_i)/(\theta_s - \theta_i)$ is the relative water content with θ_i being initial volumetric water content and θ_s being saturated volumetric water content. $F(\Theta)$ is the "flux concentration relation," formulated by Philip and Knight (21), defined by

$$F(\Theta) = \frac{q - K(\theta_i)}{R - K(\theta_i)} \tag{5.27}$$

where q is the soil water flux in the soil.

Zhu et al. (19) demonstrated, through a numerical analysis of Equation (5.26), the dependence of ponding time on slope inclination. Figure 5.30 shows t_p versus degree of a slope covered with Yolo light clay, whose saturated hydraulic conductivity is $0.04\,\mathrm{cm\,h^{-1}}$, under very small rainfall intensities

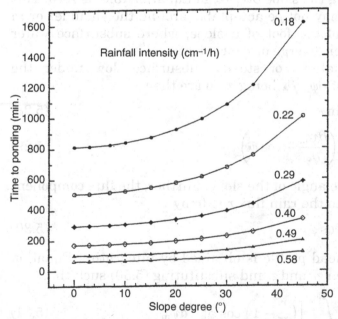

Figure 5.30 Ponding time as a function of slope degree on Yolo light clay with initial volumetric water content $0.20\,\mathrm{cm^3\,cm^{-3}}$. (After Zhu, D., Nakano, M., and Miyazaki, T., Numerical simulation of time to ponding and water flow in slope, *Trans. Jpn. Soc. Irrig. Drain. Reclam. Eng. 194*:73–80 (1998). With permission.)

that are still exceeding this saturated hydraulic conductivity. As is supposed from the infiltration capacity concept on slopes shown in Figure 5.21, the steeper the slopes, the larger the t_p values, especially under very low rainfall intensities.

V. SUBSURFACE FLOW OF WATER IN SLOPES

A. One-Dimensional Steady Flow in Slopes

1. Water Flow in Uniform Slopes

Many reports have pointed out both experimentally and theoretically that equipotential lines of the matric head in slopes are mostly parallel to each slope during infiltration, percolation, and drainage of water. Hence, it is reasonable to analyze the subsurface flow of water in slopes using Equations (5.6) and (5.7) with the assumption that the matric head gradient along the slope $\partial \psi_m / \partial x$ in Equation (5.6) is zero. This assumption may not be acceptable around the shoulder or in the vicinity of the foot of a slope, where subsurface water tends to either diverge or concentrate.

The equations of steady subsurface flow under the assumption of $\partial \psi_m / \partial x$ being zero are then

$$q_x = K \sin \phi \qquad (5.28)$$

$$q_z = -K \left(\frac{\partial \psi_m}{\partial z} + \cos \phi \right) \qquad (5.29)$$

where ϕ is the angle of the slope. Further, the flux component q_z is related to the rain flux r (>0) by

$$q_z = -r \cos \phi \qquad (5.30)$$

The matric head profile is obtained by integrating Equation (5.29) between z_0 and z and substituting (5.30) such that

$$z - z_0 = \int_{\psi_{m0}}^{\psi_m} \left[\left(\frac{r}{K} - 1 \right) \cos \phi \right]^{-1} d\psi_m \qquad (5.31)$$

where z_0 is the reference height at which the matric head is ψ_{m0}. Equation (5.31) has been solved for the case of ϕ being zero by several researchers, as mentioned in Chapter 2.

Miyazaki (22) solved Equation (5.31) for the different values of ϕ by using an approximated unsaturated hydraulic conductivity of sandy loam given by Equation (5.17). A curve of K versus ψ_m is denoted in Figure 5.16 using experimental values. Integrating Equation (5.31) with Equation (5.17), he obtained

$$z - z_0 = \frac{12.5}{r\cos\phi}\frac{1}{4a^3}\left[\ln\left|\frac{\psi_m + a}{\psi_m - a}\right| - 2\arctan\frac{-\psi_m}{a}\right]_{\psi_{m0}}^{\psi_m}$$

(5.32)

where

$$a = \left(\frac{12.5}{r} - 4460\right)^{0.25}$$

(5.33)

Figure 5.31 shows a rectangular slope-adjustable lysimeter 50-cm long, 30-cm deep, and 50-cm wide filled with this sandy loam. The reference height z_0 is zero along the bottom of the lysimeter.

Figure 5.32 shows the theoretical (solid lines) and experimental (dots) profiles of (a) matric head $-\psi_m$, (b) volumetric water content θ, and (c) lateral water flux q_x in a midslope of

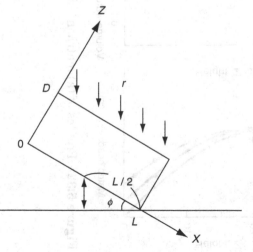

Figure 5.31 Sizes of a rectangular slope-adjustable lysimeter.

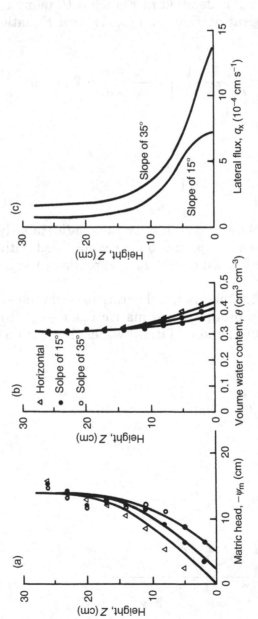

Figure 5.32 Profiles of (a) matric head, (b) volume water content, and (c) lateral flux in various slopes.

$0°$, $15°$, and $35°$ under a constant rainfall of $10\,\text{mm}\,\text{h}^{-1}$. In the calculation, the boundary condition of matric head ψ_{m0} at the midpoint of the bottom of the lysimeter was given by

$$\psi_{m0} = -\frac{L}{2}\sin\phi \tag{5.34}$$

The theoretical water contents in Figure 5.32(b) were converted from a theoretical solution of ψ_m using Equation (5.15). The theoretical lateral water fluxes in Figure 5.32(c) were obtained by substituting the theoretical solutions of ψ_m into Equation (5.17) and then substituting the estimated values of K into Equation (5.28). It is obvious from Figure 5.32 that the integration technique given by Equation (5.31) is applicable to the one-dimensional analysis of steady rain percolation in midslopes.

2. Water Flow in Layered Slopes

In natural slopes, soils are almost always stratified due to the colluviation, accumulation, and weathering of soils. Figure 5.33 shows an example of a layered slope observed at an outcrop in Tochigi prefecture in Japan. The interfaces

Figure 5.33 Outcrop of a layered slope.

between adjacent layers are very clear and the physical property of each layer is markedly different. The effects of such layers in natural slopes as exemplified here on the flow of water, however, have been vague.

Steady flow of water in a layered slope is also analyzed by means of the integration technique. A schematic diagram of soil layers parallel with a given slope is shown in Figure 5.34, where the saturated or unsaturated hydraulic conductivities of each layer were defined independently. Under a steady state, the components of flux in the direction of the z-axis q_z value are always maintained constant according to the refraction law (see Chapter 3) and are given by Equation (5.30). Assuming that each layer is parallel with the slope, Equation (5.29) is integrated at each layer and a continuous matric head profile is obtained.

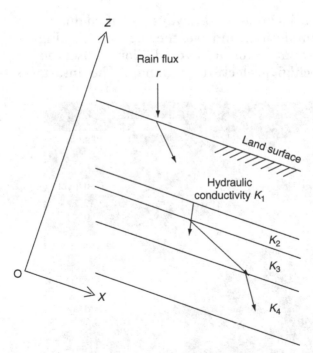

Figure 5.34 Soil layers parallel to a slope and steady fluxes of water in each layer.

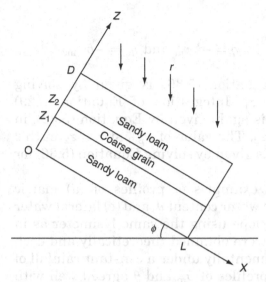

Figure 5.35 Sizes of a rectangular slope-adjustable lysimeter sandwiching a coarse grain layer within sandy loam.

Miyazaki (22) investigated matric head profiles, both experimentally and theoretically, under a steady rainfall in a slope sandwiching a very coarse grained layer within sandy loam layers, as shown in Figure 5.35, where z_1 and z_2 denote the z distances of the limits of the very coarse grained layer. The saturated hydraulic conductivity of the very coarse grained layer was $0.91 \, \text{cm s}^{-1}$, and its unsaturated hydraulic conductivity was assumed to be

$$k = \frac{0.91}{(-\psi_m)^6} \tag{5.35}$$

which is shown in Figure 5.16. Integration of Equation (5.29) from zero to z ($0 \leq z \leq z_1$) is given by Equation (5.32), in which z_0 is zero, and integration from z_1 to z ($z_1 \leq z \leq z_2$) is given by

$$z - z_1 = \frac{1}{b \cos \phi} \left[\frac{1}{12} \ln \left| \frac{\eta^2 - \eta + 1}{\eta^2 + \eta + 1} \right| + \frac{1}{6} \ln \left| \frac{\eta - 1}{\eta + 1} \right| \right.$$
$$\left. + \frac{1}{3.464} \arctan \frac{1.732 \eta}{1 - \eta^2} \right]_{\eta_0}^{\eta} \tag{5.36}$$

where

$$b = \left(\frac{r}{0.91}\right)^{1/6}, \quad \eta = -b\psi_m \text{ and } \eta_0 = -b\psi_{m0}$$

The value of ψ_{m0} in Equation (5.36) is given by solving Equation (5.32) for $z = z_1$. Integration of Equation (5.29) from z_2 to z ($z_2 \leq z \leq D$) is again given by Equation (5.32), in which z_0 is replaced by z_2. The value of ψ_{m0} at $z = z_2$ in this integration from z_2 to D is given by solving Equation (5.36) for $z = z_2$.

Figure 5.36 shows examples of profiles of (a) matric head $-\psi_m$, (b) volumetric water content θ, and (c) lateral water flux q_x in a three-layer slope using the same lysimeter as in Figure 5.32. Solid lines were obtained theoretically and each plot was obtained experimentally under a constant rainfall of $10\,\mathrm{mm\,h^{-1}}$. Theoretical profiles of ψ_m and θ agreed well with experimental values, but theoretical profiles of q_x overestimated the experimental values (not presented here) of lateral flow. This discrepancy may be attributed to a preferential flow in the sandwiched very coarse grained layer, and is discussed later.

B.   Two-Dimensional Behavior of Water in Slopes

1.   Distribution of Flux

One-dimensional analysis of flow is applicable to the midslope but not necessarily to the top, shoulder, foot, or bottom of slopes, as the boundary conditions at the top and bottom of slopes influence the flow. Miyazaki (23) measured the matric head distribution in a layer-model slope sandwiching a plant layer during a steady percolation of water. He denoted the flux distribution in the slope as shown in Figure 5.37, where the quantities and directions of the flow at every site in the slope are given by the lengths and directions of arrows. The lateral component of flow was largest just above the interface between the topsoil and the second plant layer, due to the capillary barrier effect mentioned in Chapter 4. Evidently, the flux distribution at the upper site, where only a little flux exists in the third soil layer, is different from that at

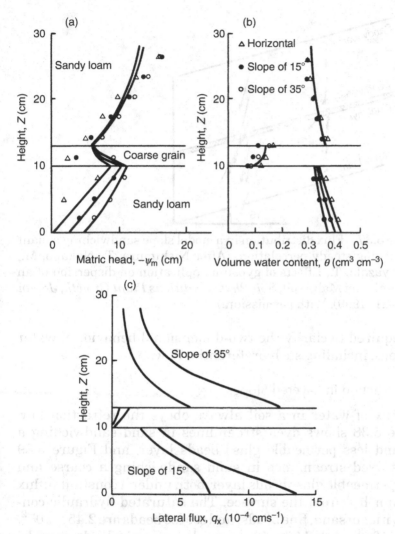

Figure 5.36 Profiles of (a) matric head, (b) volume water content, and (c) lateral flux in various layered slopes.

the lower site, where larger fluxes exist in both the top and third soil layers.

The flux distributions in a layered slope, obtained experimentally (Figure 5.37), are affected by preferential flow such as that of a capillary barrier or funneled flow along the interface of a soil layer. Further theoretical and experimental studies

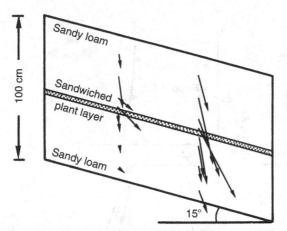

Figure 5.37 Flux distribution in a model slope sandwiching a plant layer during steady percolation. (After Nishimura, T., Nakano, M., and Miyazaki, T., Effects of gypsum application on dispersion of an acid Kunigami Mahji soil, *Soil Phys Conditions Plant Growth, Japan 81*: 15–21 (1999). With permission.)

are required to clarify the two-dimensional behavior of water in slopes, including such preferential flows.

2. Refraction in Layered Slopes

The flux of water in a soil always obeys the refraction law. Figure 5.38 shows dyed streamlines in sand sandwiching a fine and less permeable glass beads layer, and Figure 5.39 shows dyed streamlines in sand sandwiching a coarse and more permeable glass beads layer, both under a constant influx of $50 \, mm \, h^{-1}$ from the surface. The saturated hydraulic conductivities of sand, fine beads, and coarse beads are 2.45×10^{-2}, 3.09×10^{-3}, and $6.86 \times 10^{-1} \, cm \, s^{-1}$, respectively. It may be striking that when the downward flow of water in sand encounters a finer, less permeable layer, no lateral flow exists in sand and the flux refracts at the boundary inside the slope (Figure 5.38), whereas when the flow encounters a coarser and more permeable layer, the flux bends in the down slope direction, causing lateral water flow in the sand layer (Figure 5.39).

These illustrations apparently contradict general belief, which is that lateral water flow occurs along less permeable

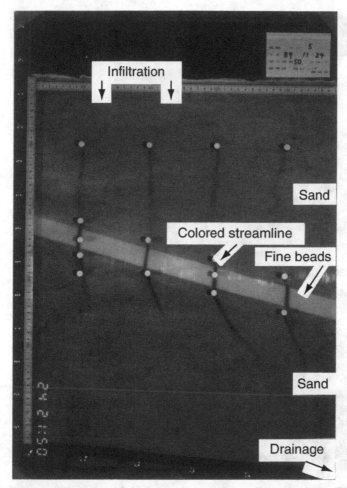

Figure 5.38 Dyed streamlines in sand sandwiching a less permeable glass beads layer.

subsoils in slopes. The phenomenon shown here is, however, quite reasonable according to the refraction law of unsaturated soils (see Chapter 3),

$$\frac{K_1}{K_2} = \frac{\tan \alpha}{\tan \beta} \tag{3.4}$$

where K_1 and K_2 are the unsaturated hydraulic conductivities of the top, sand layer and the second, glass beads layer. In

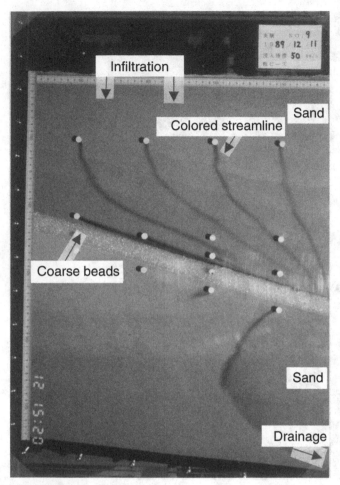

Figure 5.39 Dyed streamlines in sand sandwiching a more permeable glass beads layer.

Figure 5.38, since sand is more permeable than fine beads, α must be larger than β. On the other hand, in Figure 5.39, since sand is much more permeable than coarse beads in an unsaturated condition, α must be much larger than β. It should be noted that large pores generally have very small permeabilities in unsaturated soils.

Figure 5.40 gives typical refraction flows in slopes where K_1 is the hydraulic conductivity on the upstream side and K_2 is the hydraulic conductivity on the downstream side. When

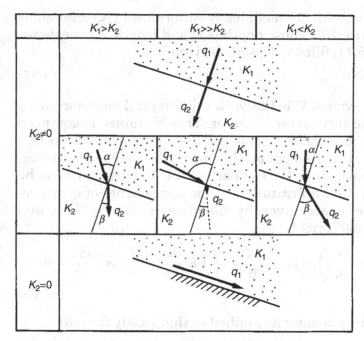

Figure 5.40 Typical refraction flows in slopes.

the interface between two layers lies on a total head contour, no refraction occurs at the interface, because both incidence and refraction water fluxes must be perpendicular to the interface. When K_1 is larger than K_2, the incidence angles α are larger than β, and when K_1 is smaller than K_2, α is smaller than β. The flow on an impermeable sloping bed is regarded as the case where K_2 is infinitely small, resulting in α being 90°.

3. Mathematical Models

Mathematical investigations of two-dimensional water flow in slopes enable us to understand the flow in slopes more realistically. Readers who are interested in the numerical analysis of two-dimensional water flow in slopes may refer to specialized books for details. A simple numerical approximation of two-dimensional water flow is given here to estimate the effects on the flow of boundary conditions on the upper and lower ends of a slope.

The basic equation of steady water flow in a rectangular container $AA'BB'$, whose depth is D and length is L as shown in Figure 5.41, filled with soil, is given by

$$\nabla \cdot \boldsymbol{q} = 0 \tag{5.37}$$

where the symbol ∇ is the vector differential operator and $\boldsymbol{q}$ denotes the flux vector of water. This container is equipped with an impermeable left sidewall and bottom, but allows free water and air movement on the upper and right sides. Hence, water is soaked from the soil surface AA' and drained from the lower end, $A'B'$. Substitution of the components of $\boldsymbol{q}$ in two-dimensional space, given by Equations (5.6) and (5.7), into Equation (5.37) yields

$$\frac{\partial}{\partial x}\left(K\frac{\partial \psi_m}{\partial x}\right) - \sin\phi\frac{\partial K}{\partial x} + \frac{\partial}{\partial z}\left(K\frac{\partial \psi_m}{\partial z}\right) + \cos\phi\frac{\partial K}{\partial z} = 0$$

$$\tag{5.38}$$

The boundary conditions applied to this steady flow are

$$r = K, \ 0 \le x \le L, \ z = D \tag{5.39}$$

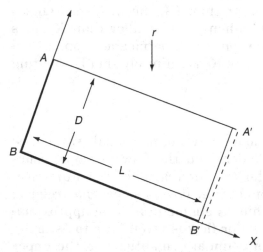

Figure 5.41 Boundary conditions of a rectangular container.

for the soil surface AA',

$$\frac{\partial \psi_m}{\partial z} + \cos \phi = 0, \quad 0 \le x \le L, \quad z = 0 \tag{5.40}$$

for the bottom BB',

$$\frac{\partial \psi_m}{\partial x} - \sin \phi = 0, \quad x = 0, \quad 0 \le z \le D \tag{5.41}$$

for the upper end of the slope AB, and

$$\psi_m = 0, \quad x = L + \Delta x, \quad 0 \le z \le D \tag{5.42}$$

for the lower end of the slope $A'B'$ where Δx is a very small distance from the right-hand side, $A'B'$. There are many other ways of stating the boundary conditions for the problem, especially for the free seepage surface $A'B'$, which may provide slightly different approximations of the solution to Equation (5.38).

Figure 5.42 shows numerical solutions of matric head distributions in the model, subjected to the following conditions:

Rain intensity	$r = 20 \, \text{mm} \, \text{h}^{-1}$
Slope angles	$\phi = 15°$ and $35°$
Length	$L = 250 \, \text{cm}$
Depth	$D = 28 \, \text{cm}$

Matric head profiles in the direction of the z-axis on 50-cm intervals of the x-axis are denoted in these figures. The lines rectangular to the slopes at the 50-cm intervals represent the z-axis at every point on the slopes, and the points where the matric head profiles cross the z-axis are the groundwater levels. The groundwater level can thus be drawn by combining the cross points for both slopes. Many publications deal with numerical solution of water flow in slopes, and readers are referred to the books by Remson et al. (24) and Huyakorn and Pinder (25) for details.

Figure 5.43 shows the matric head profiles at every 25 cm on the x-axis of the slopes. The profiles are almost similar in midslope, as expected, but differ near the upper and lower

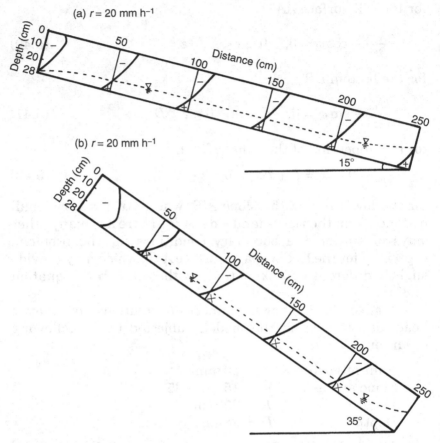

Figure 5.42 Numerical solutions of matric head distributions in model slopes.

ends of the slopes. By investigating these discrepancies, the distance at which matric head profiles are affected by the sidewalls of slopes is estimated. In this example, the distance of influence is estimated at about 25 cm from the upper and lower ends of the 15° slope (Figure 5.43a) and about 50 cm from the ends of the 35° slope (Figure 5.43b). The distance at which the upper and lower boundary conditions of a model slope influence the matric head distributions

(a) Slope of 15°

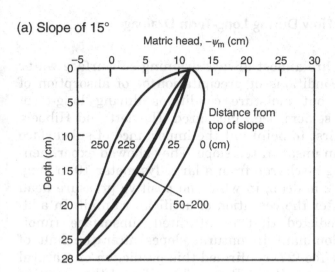

(b) Slope of 35°

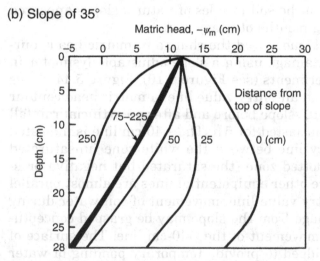

Figure 5.43 Matric head distributions at every site of two slopes.

depends on the rain intensity, angle of slope, soil thickness, and hydraulic properties of the soils. Generally, it is probably safe to say that the thinner the soil layers and the smaller the angles of inclinations, the less the distance of influence of the upper and lower boundary conditions of slopes.

C. Subsurface Flow During Long-Term Drainage in Slopes

Natural slopes are almost always draining subsurface water except under conditions of precipitation or of absorption of melting snow, but moisture conditions during long-term drainage have seldom been measured. Hewlett and Hibbert (26) were the first to point out the importance of subsurface water flow in an unsaturated slope. They showed experimentally the lasting discharge from a large lysimeter 10-m long, 2-m wide, and 2-m deep, in which no positive pressure head zone existed after the cessation of artificial rainfall. In addition, they predicted that unsaturated subsurface runoff would be predominant in natural slopes after the halt of rain. Harr and Yee (27) confirmed this prediction on natural slopes in the Oregon Coast Range. They found that positive water pressure in the soil profiles of natural slopes appeared only once during months of heavy rain.

Miyazaki (1) measured the change in matric heads during long-term drainage using a slope-adjustable lysimeter in infiltration experiments (see Figure 5.10). Figure 5.44 shows the locations of an air entry value line (a matric head contour of -40 cm) in a $10°$ slope before and after an artificial rainfall of $20\,\mathrm{mm\,h^{-1}}$ that lasted for 5 h. The -40-cm line is indicated by the boundary line between the white zone (unsaturated zone) and the dotted zone (the saturated but negative pressure zone). Since other equipotential lines are almost parallel with the air entry value line, movement of soil water during long-term drainage from the slope may be grasped conceptually by tracing movement of the -40-cm line. The surface of the slope was ridged to provide temporary ponding of water during artificial rain conditions.

Figure 5.44(a) shows the initial condition of the -40-cm line. One hour following cessation of rain, the matric head for the entire slope was between 0 and -40 cm. At the bottom of the slope, groundwater (the saturated and positive pressure zone) appeared, as indicated by the dark zone in Figure 5.44(b). One day after termination of the rain, the groundwater disappeared and the -40-cm line was located near the

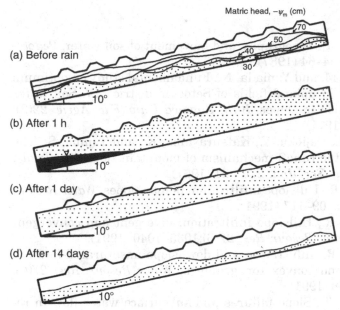

Matric head, $-\psi_m$ (cm)

(a) Before rain

(b) After 1 h

(c) After 1 day

(d) After 14 days

Figure 5.44 Location of equipotential lines of air entry value (a) before a rainfall and (b–d) after a rainfall.

soil surface, parallel with the slope, as shown in Figure 5.44(c). Figure 5.44(d) shows that the −40-cm line was still parallel to the slope 14 days after the end of the rain. Discharge from the drain pipe installed at the bottom edge of the lysimeter continued for more than 14 days after halting the artificial rainfall. It is notable that discharge continued even after the groundwater disappeared and that the entire slope, except for the bottom edge of the lysimeter, was unsaturated. This result agrees well with the famous experiment by Hewlett and Hibbert (26) mentioned above.

It is supposed from the discussions above that equipotential lines will be almost parallel to the slope during long-term subsurface drainage, except for very soon after the half of rainfall, when the phreatic surface appears in the vicinity of the bottom of the lysimeter. The depth of the impermeable layer in natural soils may influence downward advancement of the equipotential lines during drainage.

REFERENCES

1. Miyazaki, T., Topography and movement of soil water, *Pedologist 31*(2):54–64 (1987).
2. Matsuda, M. and Yamada, N., Fundamental study on optimum irrigation in upland fields of Setouchi district. IV. Soil water consumption in slopes, *Bull. Kagawa Univ. Fac. Agric. 22*(2): 113–117 (1971).
3. Ooeda, M., Fujioka, Y., Katsurayama, K., and Tajima, S., The study on the runoff mechanism of mountain slopes (I), *Trans. Agric. Eng. Soc. Jpn. 2*:75–81 (1961).
4. Philip, J. R., Hillslope infiltration: planar slopes, *Water Resour. Res. 27*(1):109–117 (1991).
5. Philip, J. R., Hillslope infiltration: divergent and convergent slopes, *Water Resour. Res. 27*(6):1035–1040 (1991).
6. Philip, J. R., Infiltration and downslope unsaturated flows in concave and convex topographies, *Water Resour. Res. 27*(6): 1041–1048 (1991).
7. Miyazaki, T., Slope failures and subsurface water flow in reclaimed farm lands, *Jpn. Agric. Res. Q. 20*(3):174–179 (1987).
8. Redinger, G. J., Campbell, G. S., Saxton, K. E., and Papendick, R. I., Infiltration rate of slot mulches: measurement and numerical simulation, *Soil Sci. Soc. Am. J. 48*:982–986 (1984).
9. Miyazaki, T., Water infiltration into layered soil slopes, *Trans. Jpn. Soc. Irrig. Drain. Reclam. Eng. 133*:1–9 (1988).
10. Miller, D. P. and Gardner, W. H., Water infiltration into stratified soil, *Soil Sci. Soc. Am. Proc. 26*:115–119 (1962).
11. Hillel, D., *Application of Soil Physics,* Academic Press, New York, p. 26 (1980).
12. Horton, R. E., The role of infiltration in the hydrologic cycle, *Trans. Am. Geophys. Union 14*:446–460 (1933).
13. Dunne, T. and Black, R. D., Partial area contributions to storm runoff in a small New England watershed, *Water Resour. Res. 6*(5):1296–1311 (1970).
14. Chorley, R. J., The hillslope hydrological cycle, in *Hillslope Hydrology*, M. J. Kirby, Ed., Wiley, Chichester, pp. 1–42 (1978).
15. Tanaka, T., Yasuhara, M., Sakai, H., and Marui, A., The Hachioji experimental basin study: storm runoff processes and the mechanism of its generation, *J. Hydrol. 102*:139–164 (1988).

16. Nishimura, T., Nakano, M., and Miyazaki, T., Effects of crust formation on soil erodibility, *Trans. Jpn. Soc. Irrig. Drian. Reclam. Eng. 146*:101–107 (1990).

17. Nishimura, T., Nakano, M., and Miyazaki, T., Effects of clod size, raindrop size and initial moisture conditions on surface sealing of Kunigami Mahji soil, *Trans. Jpn. Soc. Irrig. Drain. Reclam. Eng.* 199:*17*–22 (1999).

18. Nishimura, T., Nakano, M., and Miyazaki, T., Effects of gypsum application on dispersion of an acid Kunigami Mahji soil, *Soil Phys Conditions Plant Growth, Japan 81*: 15–21 (1999).

19. Zhu, D., Nakano, M., and Miyazaki, T., Numerical simulation of time to ponding and water flow in slope, *Trans. Jpn. Soc. Irrig. Drain. Reclam. Eng. 194*:73–80 (1998).

20. Kutilek, M., Constant rainfall infiltration, *J. Hydrol. 45*:289–303 (1980).

21. Philip, J. R. and J. H. Knight, On solving the unsaturated flow equation, 3, New quasi-analytical techniques, *Soil Sci. 117*(8):1–13 (1974).

22. Miyazaki, T., Water movement in soils on slopes, *Soil Phys. Cond. Plant Growth Jpn. 49*:40–47 (1984).

23. Miyazaki, T., Water flow in unsaturated soil in layered slopes, *J. Hydrol. 102*:201–214 (1988).

24. Remson, I., Hornberger, G. M., and Molz, F. J., *Numerical Methods in Subsurface Hydrology,* Wiley-Interscience, New York (1971).

25. Huyakorn, P. S. and Pinder, G. F., *Computational Methods in Subsurface Flow,* Academic Press, London (1983).

26. Hewlett, J. D. and Hibbert, A. R., Moisture and energy conditions within a sloping soil mass during drainage, *J. Geophys. Res. 64*:1081–1087 (1963).

27. Harr, R. D. and Yee, C. S., Soil and hydrologic factors affecting the stability of natural slopes in Oregon Coast Range, *WRRI–33,* Water Resources Research Institute, Oregon State University, Corvallis, OR (1975).

6

Water Flow Under the Effects of Temperature Gradients

219

I. TEMPERATURE PROFILE AND SOIL WATER FLOW IN FIELDS

The temperature gradient of soil is a driving force for the movement of both liquid water and water vapor. Therefore, a daily change in the temperature profile may influence water movement in the soil. Figure 6.1 shows a typical daily change of wind velocity, net radiation from the atmosphere, relative humidity, and air temperature near the ground, measured simultaneously in a sand dune field in summer (1). The heights of each measurement are indicated in the figure. Figure 6.2 is the corresponding daily change of soil temperature and volumetric soil water content at each depth. Both the highest temperature and the lowest temperature appear at the land surface, being 48°C at 1:00 p.m. and 23°C at 5:00 a.m., respectively. Although the total amount of water in this soil is decreasing during 24 h of measurement, it should be noted in Figure 6.2 that there is an evident temporal increase in volumetric water content in the surface

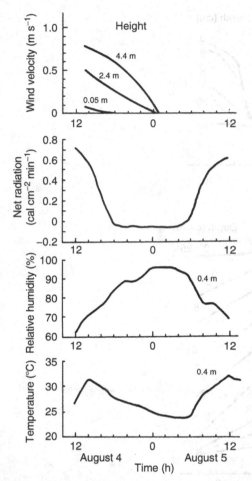

Figure 6.1 Typical daily change of wind velocity, net radiation from atmosphere, relative humidity, and air temperature in a sand dune field. (From Miyazaki, T. and Amemiya, Y., *Trans. Jpn. Soc. Irrig. Drain. Reclam. Eng. 48*:16–22 (1973). With permission.)

soil at midnight, although it disappears with sunrise. This type of temporal moisture increase in the surface zone may be preferable for plant growth in such fields under arid and semiarid weather conditions. Both liquid water and water vapor move in such a sand dune field by the driving forces of gravity, matric potential gradient, temperature gradient, and, presumably, solute concentration gradient.

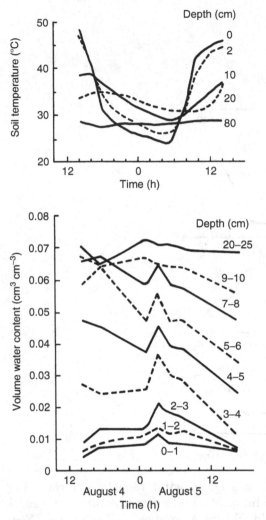

Figure 6.2 Daily change in soil temperature and volume soil water content at each depth. (From Miyazaki, T. and Amemiya, Y., *Trans. Jpn. Soc. Irrig. Drain. Reclam. Eng. 48*:16–22 (1973). With permission.)

The water balance in an REV under the effects of a temperature gradient is as shown in Figure 6.3. The equation of water balance is

$$\Delta S = (P + I + C) - E + q \tag{6.1}$$

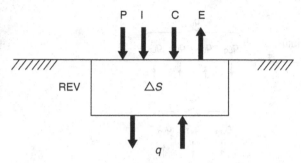

Figure 6.3 Water balance in the REV at land surface.

where ΔS is the increase of water in REV, P is the precipitation, I is the irrigation, C is the condensation from the atmosphere, E is the evaporation, and q is the flux of water in the soil. Since Equation (6.1) is a conceptual water balance equation, the dimensions of each term must be modified appropriately in the quantitative calculations. Water flux is composed of the following components:

q_L (liquid flux)
    q_{Lg} (induced by gravity)
    q_{Lm} (induced by matric potential gradient)
    q_{LT} (induced by temperature gradient)
    q_{Lo} (induced by osmotic potential gradient)
q_V (vapor flux)
    q_{Vm} (induced by matric potential gradient)
    q_{VT} (induced by temperature gradient)
    q_{Vo} (induced by osmotic potential gradient)

Flux q in Equation (6.1) is positive when it flows into the REV and is negative when it flows out of the REV.

Figure 6.4 shows conceptual profiles of temperature T, matric potential ϕ_m, and osmotic potential ϕ_o in semiarid or arid regions, where annual evaporation exceeds annual precipitation. In the daytime, the directions of thermally driven fluxes q_{LT} and q_{VT} are downward, due to the temperature gradient near the land surface (Figure 6.4a). At night, however, as the temperature profile decreases upward, q_{LT} and q_{VT} may turn up (Figure 6.4b). Since annual evaporation exceeds the annual precipitation here, salt accumulation at

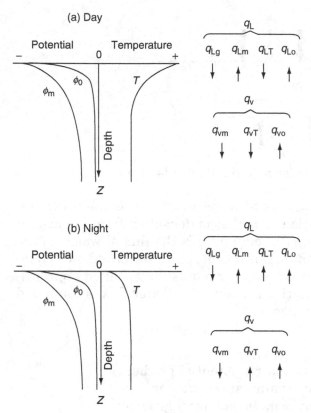

Figure 6.4 Profiles of temperature, matric potential, and osmotic potential in semiarid or arid regions (a) during the day and (b) at night, with resultant water flux directions.

the land surface generally increases with time, resulting in a reduction of osmotic potential ϕ_0 and in upward fluxes q_{Lo} and q_{Vo}. If annual precipitation exceeds annual evaporation, there may only be a slight gradient of osmotic potential near the land surface due to natural leaching. The directions of other components of q_L and q_V are designated in Figure 6.4 by arrows.

Brawand and Kohnke (2) measured the annually integrated amount of condensation of water vapor from the atmosphere (C in Equation (6.1)) and from the lower soil layer (q_V) in the field, separately. They estimated experimentally that 31 mm of water vapor condensed from the atmosphere to the land surface in 1 year, while 34 mm of water vapor moved upward in the soil and condensed within the top soil layer.

A physically based estimation of each flux given above will be useful in understanding the influence of the temperature profile on soil water flow in fields. Generally speaking, however, quantitative predictions of each flux described above still have practical difficulties, due mainly to the difficulty in determining the parameters included in the equations for each flux. It is much more difficult to separate the measurements for each flux.

II. WATER VAPOR FLOW UNDER A TEMPERATURE GRADIENT

A. Simple Model

1. Isothermal Diffusion of Water Vapor in Soils

Figure 6.5(a) is a closed space divided by a vertical plane by which dry air and water vapor, both under the same total pressure P and the same temperature T, are separated. When the vertical plane is removed smoothly, the two gases will diffuse as illustrated in Figure 6.5(b), conforming to Fick's law:

$$q_v^* = -D\frac{d\rho_v}{dx} \tag{6.2}$$

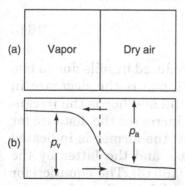

Figure 6.5 Dry air and water vapor (a) in closed spaces and (b) their mutual diffusion.

$$q_a^* = -D \frac{d\rho_a}{dx} \tag{6.3}$$

where q_v^* and q_a^* are the diffusion fluxes of water vapor and dry air in a space (kg s^{-1} m^{-2}), respectively; ρ_v and ρ_a are the densities of water vapor and dry air (kg m^{-3}), respectively; D is the mutual diffusion coefficient identical for both water vapor and dry air (m^2 s^{-1}); and x is the distance (m). The dimensions of fluxes q_v^* and q_a^* are transformed into meters per second by dividing the right-hand sides of Equations (6.2) and (6.3) by the density of water (1 Mg m^{-3}).

Assuming that water vapor and dry air are ideal gases, the state equations

$$p_v = \rho_v R_v T \tag{6.4}$$

$$p_a = \rho_a R_a T \tag{6.5}$$

are used, where p_v and p_a are the pressure fractions of water vapor and dry air (N m^{-2}), respectively; R_v and R_a are the gas constants of water vapor and dry air (J kg^{-1} K^{-1}), respectively; and T is the absolute temperature (K). The gas constants R_v and R_a are determined by dividing the molecular gas constant 8.317 J mol^{-1} K^{-1}, which is independent of both temperature and type of gas, by the molecular weight of water vapor, 0.018 kg mol^{-1}, and that of dry air, 0.029 kg mol^{-1}, such that R_v is 462.1 J kg^{-1} K^{-1} and R_a is 286.8 J kg^{-1} K^{-1}. Substituting Equation (6.4) into Equation (6.2), we obtain

$$q_v^* = -\frac{D}{R_v T} \frac{dp_v}{dx} \tag{6.6}$$

The diffusion of water vapor is reduced in soils due to two factors, as illustrated in Figure 6.6; one is the decrease in space where gas can pass through, and another is the irregularities of pore spaces in soils, which increases the distance for the movement of gas. Quantitatively, the former is indicated by the volume ratio of the gas phase and the latter by the tortuosity, whose value is generally 0.66 (3). The equation for water vapor flux is then given by multiplying these factors by Equation (6.6),

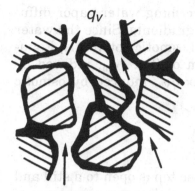

Figure 6.6 Simple model of water vapor diffusion in soil pores.

$$q_V = -\frac{Da\xi}{R_v T}\frac{dp_v}{dx} \tag{6.7}$$

where q_V is the flux of water vapor in the soil, a is the volume fraction of gas phase in the soil, and ξ is the tortuosity. According to the definition of equilibrium, the relative humidity (RH) of air is given using the water potential ϕ_w as

$$RH = \frac{p_v}{p_0} = \exp\left(\frac{\phi_w}{R_v T}\right) \tag{6.8}$$

and hence

$$p_v = p_0 \exp\left(\frac{\phi_w}{R_v T}\right) \tag{6.9}$$

where p_0 is the saturated water vapor pressure at a given temperature. Note that the water potential ϕ_w is the sum of the matric potential and the osmotic potential ϕ_o of soil water,

$$\phi_w = \phi_m + \phi_o \tag{6.10}$$

in joules per kilogram. Substituting Equation (6.9) into Equation (6.7), we obtain

$$q_V = -\frac{Da\zeta p_v}{(R_v T)^2}\frac{d\phi_w}{dx} \tag{6.11}$$

Equation (6.11) is simplest for describing water vapor diffusion in a soil without temperature gradients. Since the water potential consists of the matric and osmotic potentials, Equation (6.11) is regarded as the sum of flux q_{Vm}, induced by matric potential gradient, and flux q_{Vo}, induced by osmotic potential gradient.

2. Enhancement Factor by One-Way Diffusion in Soils

When water is ponded in a pit whose top is open to a still and dry atmosphere as shown in Figure 6.7, water vapor moves upward due to the vapor pressure gradient, whereas air in the pit does not move downward despite the existence of an air pressure gradient. This is attributed to the cancellation of air diffusion and convection. Under these conditions the upward movement of water vapor is enhanced as described below.

Instead of Equations (6.2) and (6.3), we use the flux equations

$$q_v^* = -D\frac{d\rho_v}{dz} + s\rho_v \tag{6.12}$$

$$q_a^* = -D\frac{d\rho_a}{dz} + s\rho_a \tag{6.13}$$

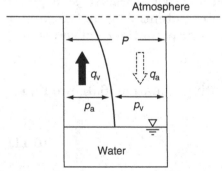

Figure 6.7   Water vapor diffusion from ponded water in a pit.

where the second terms of both equations give convection fluxes and s is the velocity of convection (m s^{-1}). When the air flux q_a^* is zero, the velocity of convection is given by

$$s = \frac{D}{\rho_a} \frac{d\rho_a}{dz} \qquad (6.14)$$

Taking account of the fact that the total pressure P, given by

$$P = p_v + p_a \qquad (6.15)$$

is constant, together with the ideal gas assumptions (Equations (6.4) and (6.5)), we obtain a one-way diffusion equation for water vapor:

$$q_v^* = -\frac{D}{R_v T} \frac{P}{P - p_v} \frac{dp_v}{dz} \qquad (6.16)$$

Since the term $P/(P - p_v)$ is always larger than 1, this term is an enhancement factor due to one-way diffusion. Rollins et al. (4) termed this factor the mass flow factor.

Generally, when the transient flow of liquid water is small enough, air in soil pores is assumed not to move, due to the existence of groundwater or an impermeable layer in the bottom. Under such conditions, the equation of one-way diffusion of water vapor should be applied in soils rather than equations of mutual diffusion (Equation (6.7) or (6.11)). Multiplying soil factor a, the volume fraction of gas, and ξ, the tortuosity, by Equation (6.16), the one-way diffusion equation for water vapor in soils is given by

$$q_V = -\frac{Da\xi\gamma}{R_v T} \frac{dp_v}{dz} \qquad (6.17)$$

or

$$q_V = -\frac{Da\xi\, p_v\gamma}{(R_v T)^2} \frac{d\phi_w}{dz} \qquad (6.18)$$

where γ is the enhancement factor (mass flow factor),

$$\gamma = \frac{P}{P - p_v} \qquad (6.19)$$

The value of γ varies naturally from 1 to about 1.08, depending on the humidity and temperature of air in soils (see Table 6.1).

3. Nonisothermal Diffusion of Water Vapor in Soils

Nakano and Miyazaki (5) proved theoretically that Equation (6.16) is exactly applicable to the nonisothermal diffusion of water vapor in air space by making only one assumption — that the total pressure P is constant, which is almost always true in nature. Inevitably, Equation (6.17) is applicable to soil pores under temperature gradients when P is constant. They showed that Equation (6.17) is developed further under a temperature gradient.

Under a temperature gradient, the pressure of water vapor in soil pores and its total differential are given as functions of both temperature and water potential, such that

$$p_{\mathrm{v}} = p_{\mathrm{v}}(T, \phi_{\mathrm{w}}) \tag{6.20}$$

$$\mathrm{d}p_{\mathrm{v}} = \frac{\partial p_{\mathrm{v}}}{\partial T}\,\mathrm{d}T + \frac{\partial p_{\mathrm{v}}}{\partial \phi_{\mathrm{w}}}\,\mathrm{d}\phi_{\mathrm{w}} \tag{6.21}$$

By using Equation (6.21), Equation (6.17) is developed into

$$q_{\mathrm{V}} = -\frac{\lambda Da\xi}{R_{\mathrm{v}}T}\left(\frac{\partial p_{\mathrm{v}}}{\partial T}\frac{\mathrm{d}T}{\mathrm{d}z} + \frac{\partial p_{\mathrm{v}}}{\partial \phi_{\mathrm{w}}}\frac{\mathrm{d}\phi_{\mathrm{w}}}{\mathrm{d}z}\right) \tag{6.22}$$

which is a basic equation of water vapor diffusion in soil under a temperature gradient. By introducing several physicochemical equations, Equation (6.22) is transformed into a more practical equation. The procedure for transformation of Equation (6.22) is given below.

Differentiation of the pressure fraction p_{v}, defined by Equation (6.9), with respect to T yields

$$\frac{\partial p_{\mathrm{v}}}{\partial T} = p_{\mathrm{v}}\left(\frac{1}{p_0}\frac{\partial p_0}{\partial T} - \frac{\phi_{\mathrm{w}}}{R_{\mathrm{v}}T^2}\right) \tag{6.23}$$

Nakano and Miyazaki (5) showed quantitatively that the magnitude of the first term in the right-hand side of Equation

(6.23) is extremely larger than that of the second term in the range between $-50°C$ and $+50°C$. This yields

$$\frac{1}{p_0}\frac{\partial p_0}{\partial T} >> \frac{\phi_w}{R_v T^2}$$ (6.24)

The well-known Clausius–Clapyron equation

$$\frac{\partial \ln p_0}{\partial T} = \frac{H}{R_v T^2}$$ (6.25)

gives an additional transformation of Equation (6.23), where H is latent heat of evaporation (enthalpy, J kg^{-1}). Note that the dimensions of both sides of Equation (6.25) are K^{-1}. Substituting Equations (6.24) and (6.25) into the right-hand side of Equation (6.23), we obtain a very simple equation,

$$\frac{\partial p_v}{\partial T} = \frac{p_v H}{R_v T^2}$$ (6.26)

On the other hand, partial differentiation of Equation (6.9) with respect to ϕ_w yields

$$\frac{\partial p_v}{\partial \phi_w} = \frac{p_v}{R_v T}$$ (6.27)

Substituting Equations (6.26) and (6.27) into Equation (6.22), we obtain

$$q_v = -D_s\left(\frac{H}{T}\frac{dT}{dz} + \frac{d\phi_w}{dz}\right)$$ (6.28)

where

$$D_s = \frac{\gamma Da \xi p_v}{(R_v T)^2}$$

Equation (6.28) is the most simplified expression of water vapor diffusion in soil under a temperature gradient. Further discussion of the derivation of Equation (6.28) and modifications of this equation appear in the work by Nakano and Miyazaki (5). Note that these results were obtained merely by developing Equation (6.17) under the assumption that water vapor and air behave as ideal gases.

It should also be noted that Equation (6.28), the diffusion flux of water vapor in a nonisothermal soil, is composed of the sum of three terms: the term q_{VT}, induced by temperature gradient

$$q_{VT} = -D_s \frac{H}{T} \frac{dT}{dz} \qquad (6.29)$$

the term q_{Vm}, induced by a matric potential gradient

$$q_{Vm} = -D_s \frac{d\phi_m}{dz} \qquad (6.30)$$

and the term q_{Vo}, induced by an osmotic potential gradient

$$q_{Vo} = -D_s \frac{d\phi_o}{dz} \qquad (6.31)$$

and that q_V is equal to $q_{VT} + q_{Vm} + q_{Vo}$. The latter two equations are combined into one term by using the definition of water potential (Equation (6.10)). When there is no temperature gradient, Equation (6.28) is identical to Equation (6.18). In nonisothermal soils in fields, q_{VT} predominates in dry soils, while q_{Vo} predominates in salt-affected dry soils. In nonisothermal wet soils in fields, liquid water flows surpass water vapor flows.

The most simplified expression of the thermally induced water vapor flux q_{VT} given by Equation (6.29) is not necessarily "simple" for practical use of this equation, even in a rough estimation of the flux. To obtain the approximate values of q_{VT} in a soil under a temperature gradient, it will be convenient to know the values of $D_s H/T$ in Equation (6.29) as a function of temperature. The appendix to this chapter gives the values of this term as a function of temperature.

4. Limitation of the Simple Model

Many experimental studies reveal that Equation (6.29), the equation of water vapor flux induced by temperature gradients, underestimates the measured flux. For example, the measured fluxes were 3.6 times as great as the predicted value (6), ten times as great as the predicted value (7), and five to ten times as great as the predicted value (8). Since underestimation of flux is peculiar to water vapor diffusion in soils under temperature gradients and diffusion fluxes of other gases are well predicted by simple models, there may be a particular mechanism in water vapor flow in soils under temperature gradients. It is certainly evident that limitation of the simple model (i.e., the underestimation of experimental values) is not due to lack of knowledge of driving forces of water vapor but to an underestimation of enhancement factors in D_s in Equation (6.29). Experimental investigation of such a limitation of the simple model was reviewed in detail by Iwata et al. (9).

B. Modified Model

1. Mechanistic Enhancement Factors of Water Vapor Diffusion

The equation for thermally induced water vapor flux, Equation (6.29), has been modified by multiplying mechanical enhancement factor η in order to match the simple model of flow to experimental values. The mechanical enhancement factor is then given as the ratio of water vapor flux measured experimentally by q'_{VT} to that estimated by simple model q_{VT}, such that

$$q'_{VT} = \eta q_{VT} \tag{6.32}$$

or

$$q'_{VT} = -\eta D_s H \frac{\mathrm{d}\ln T}{\mathrm{d}z} \tag{6.33}$$

It is noteworthy that even though no enhancements have been recognized for gas phases other than water vapor in soils, mechanical enhancement factors of water vapor range from about 3 to more than 10.

The most famous enhancement factor to explain the peculiar behavior of water vapor in soils was proposed by Philip and de Vries (10). Known as a *liquid island,* it is expressed by

$$\eta = \frac{[a + f(a)\theta]\zeta}{a\xi} \tag{6.34}$$

where a is the volume fraction of air in soil, θ is the volumetric water content, ζ is the ratio of microscopic temperature gradient to macroscopic temperature gradient, ξ is the tortuosity, and $f(a)$ is a function defined by

$$f(a) = \begin{cases} 1 & \text{if } a \geq a_k \\ a/a_k & \text{if } a < a_k \end{cases} \tag{6.35}$$

where a_k is a critical air-volume fraction. When a is smaller than a_k, the continuity of liquid film will be formed, resulting in the decrease of cross-sectional area for vapor flow. Otherwise, water vapor will pass through any liquid phase. The liquid island model can explain the η value of 3 to 8, which is much better than value developed by the simple model but still underestimates the experimental values.

A similar mechanical enhancement factor was proposed by July and Letey (11) as

$$\eta = \frac{[a + f(a)\theta]\zeta\xi'}{a\xi} \tag{6.36}$$

where ξ', a separate tortuosity correction factor, is given by

$$\xi' = \frac{a + \theta(a/a_k)}{a + \theta(a/a_k)(\lambda_{\text{ev}}/\lambda_1)} \tag{6.37}$$

in which λ_1 is the thermal conductivity of liquid water and λ_{ev} is the apparent thermal conductivity of water vapor.

As has been noted by Cass et al. (8) and Iwata et al. (9), the water island model proposed by Philip and de Vries (10) gave rise to a considerable advance in the theory of water vapor flow in soils under temperature gradients, and the theoretically derived mechanical enhancement factor η has provided better prediction of the water vapor flux, although some rooms are still left for better quantitative estimation of the flux by these models. To improve the accuracy of prediction, a more elaborate investigation of condensation and evaporation processes using the water island model may be helpful.

2. Condensation and Evaporation in Water Island Model

a. Equilibrium of Phases at a Flat Surface

The number of water molecules transferred from the liquid phase to the gas phase depends on the state of the liquid and is independent of the state of the gas. The number of water molecules transferred from the gas phase to the liquid phase depends on the state of the gas. Figure 6.8 shows the balance of water molecules at the interface between the gas phase and the liquid phase. Since some of the water molecules in the gas phase are reflected at the interface between the gas phase and the liquid phase, the rate of transfer of water molecules from gas to liquid and from liquid to gas is not equal under equilibrium. Denoting the number of water molecules arriving at the interface from the gas phase to the liquid phase by I_+ (s^{-1} m^{-2}) and those from the liquid phase into the gas phase by I_- (s^{-1} m^{-2}), phase equilibrium is given by

$$I_- = \alpha I_+ \tag{6.38}$$

where α is the condensation–evaporation coefficient. The physical meaning of α is given by the difference between the rotation of water molecules in the gas and liquid phases. The value of α is generally less than 1.

The number of water vapor molecules I_+ colliding with a unit area of flat liquid surface in a unit time is given by the Herz–Knudsen equation,

$$I_+ = p_0 \left(\frac{1}{2\pi mkT} \right)^{1/2} \tag{6.39}$$

where p_0 is the equilibrium water vapor pressure with flat liquid water (N m^{-2}), m is the mass of one water molecule

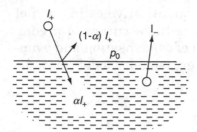

Figure 6.8 Balance of water molecules at the flat interface between the gas and liquid phases.

(kg), k is the Boltzmann constant (J K^{-1}), and T is the temperature (K). Substituting the Herz–Knudsen equation (6.39) into the equilibrium equation (6.38) and replacing k by $k = R/A$, where A is Avogadro's constant (6×10^{-23} mol^{-1}) and R is the molecular gas constant (8.317 J mol^{-1} K^{-1}), the mass basis evaporation rate mI_- (kg m^{-2} s^{-1}) is given by

$$mI_- = \alpha p_0 \left(\frac{M}{2\pi RT} \right)^{1/2} \tag{6.40}$$

where M is the molecular weight of the water (kg mol^{-1}). Since p_0 is a function of temperature only, mI_- is a function of temperature only.

b. Equilibrium at a Concave Surface

When the liquid surface is concave as shown in Figure 6.9, the mass basis evaporation rate is given by

$$mI_- = \alpha p_s \left(\frac{M}{2\pi RT} \right)^{1/2} \tag{6.41}$$

where p_s is the equilibrium water vapor pressure with concave liquid water and is lower than p_0. The ratio of p_s to p_0 is given by the Kelvin equation

$$\ln \frac{p_s}{p_0} = \frac{\sigma_w M}{\rho_w RT} \left(\frac{1}{l} - \frac{1}{d} \right) \tag{6.42}$$

where σ_w is the surface tension of liquid water, ρ_w is the density of liquid water, and l and d are the radii of curvature

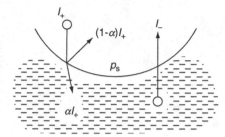

Figure 6.9 Balance of water molecules at the concave interface between the gas and liquid phases.

of the concave surface. Replacing $p_0 - p_s = \Delta p$, the left-hand side of Equation (6.42) is developed as

$$\ln\frac{p_s}{p_0} = \ln\left(1 - \frac{\Delta p}{p_0}\right)$$

$$= \left(-\frac{\Delta p}{p_0}\right) - \frac{1}{2}\left(-\frac{\Delta p}{p_0}\right)^2 + \frac{1}{6}\left(-\frac{\Delta p}{p_0}\right)^3 - \cdots \qquad (6.43)$$

When $\Delta p \ll p_0$, then

$$\ln\frac{p_s}{p_0} = -\frac{\Delta p}{p_0} \qquad (6.44)$$

Substituting Equation (6.44) into Equation (6.42), we obtain

$$p_s = p_0\left[1 + \frac{\sigma_w M}{\rho_w RT}\left(\frac{1}{l} - \frac{1}{d}\right)\right] \qquad (6.45)$$

Equation (6.41) with Equation (6.45) gives the mass basis equation of evaporation rate from a concave liquid surface.

c. Capillary Condensation at a Concave Surface

When a concave liquid surface contacts a gas phase whose vapor pressure is p_0 as shown in Figure 6.10, αI_+ will exceed the counter pass I_-, since I_+ is a function of p_0 while I_- is a function of p_s at a given temperature, resulting in an increase in the liquid phase. This phenomenon is known as *capillary condensation*.

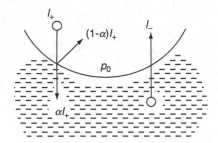

Figure 6.10 Imbalance of water molecules at the concave inter-face between the gas and liquid phases.

The rate of increase of liquid water is given by the net condensation equation

$$\alpha I_+ - I_- = \Delta p \alpha \left(\frac{M}{2\pi RT}\right)^{1/2} \qquad (6.46)$$

where

$$\Delta p = p_0 - p_s = \frac{p_0 \sigma_w M}{\rho_w RT}\left(\frac{1}{l} - \frac{1}{d}\right)$$

If a temperature gradient exists in the gas phase toward the concave interface, water vapor will diffuse toward the interface, tending to maintain the vapor pressure p_0 at the interface. The diffusion flux is given by

$$q_{VT} = -\frac{\gamma D p_v H}{(R_v T)^2}\frac{d\ln T}{dz} \qquad (6.47)$$

Thus, the vapor pressure just above the concave liquid surface is determined by the balance between the mass basis conden-sation rate αI_+ and the diffusion flux toward the liquid surface under a temperature gradient q_{VT}. Many researchers have reported that α is less than 0.1. This implies that the conden-sation rate will be smaller than the diffusion flux of water vapor when the temperature gradient is more than 1°C cm^{-1}, while it will be larger than the diffusion flux of water vapor when the temperature gradient is less than 0.01°C cm^{-1}. In other words, the continuous processes of diffusion and condensation of water vapor are governed by the rate of capillary condensation under a large temperature gradient, but they are governed by the diffusion flux of water vapor under a small temperature gradient.

d. Maximum Evaporation at a Concave Surface

The maximum evaporation rate from a concave liquid surface is given by Equation (6.41), where the gas pressure is zero. Since I_- is independent of the state of the gas phase, the mass basis evaporation rate mI_- is determined by the radii of curvature of the concave surface l and d under a given temperature. When the gas pressure is not zero, which is rather a general situation, the net evaporation rate decreases due to offset by the condensation rate αI_+, which is a function of the pressure of gas phase.

Under a temperature gradient, where the temperature is assumed to decrease with distance away from the interface as shown in Figure 6.11, the mass basis evaporation rate will be determined in such a way that the evaporation compensates the vapor flux in the direction away from the interface. Since vapor flux is proportional to the temperature gradient, it can, theoretically, increase with the temperature gradient without limit.

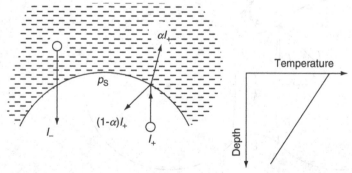

Figure 6.11 Imbalance of water molecules at the concave interface between the gas and liquid phases under a temperature gradient.

On the other hand, the maximum evaporation rate is limited by Equation (6.41), a function of temperature and curvature of the liquid surface. Therefore, if the temperature gradient is very large, the evaporation–diffusion process will be governed by the maximum evaporation rate given by Equation (6.41).

e. Steady State of Water Island

Figure 6.12 is a water island at a point of contact of two spheres under a temperature gradient, where water vapor is condensing at A and evaporating at B. Summarizing the discussion given above, the continuous processes of diffusion, condensation at A, evaporation at B, and diffusion of water vapor, are controlled by the diffusion flux under a small temperature gradient, while they are restricted by the condensation rate into the water island at A and evaporation rate from water island at B under a large temperature gradient. The governing equations are Equations (6.41), (6.46), and (6.47) for evaporation, condensation, and vapor diffusion, respectively, which are determined independent of each other.

Hence, it is concluded that when the temperature gradient is small, a steady diffusion flux exists with steady water islands in soil pores, but when the temperature gradient is

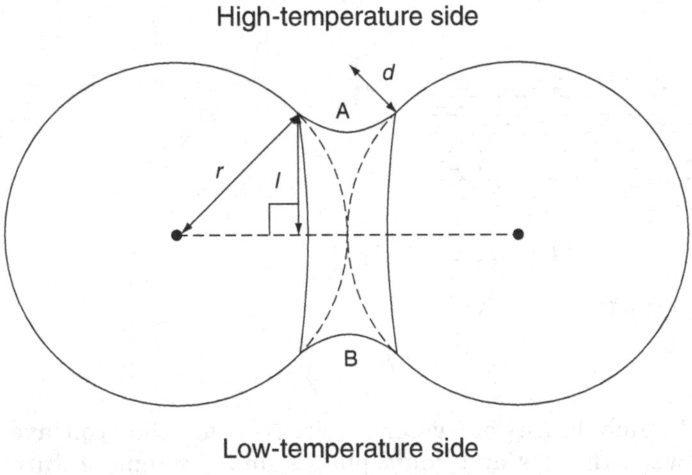

Figure 6.12   Water island at a point of contact of two spheres.

large, a transient condition will continue among vapor diffusion, capillary condensation, and evaporation processes until a renewed steady state is obtained by changing the radii of curvatures of the concave surfaces, resulting in changes in the volume of water islands.

III. LIQUID WATER FLOW UNDER A TEMPERATURE GRADIENT

A. Driving Forces Due to a Temperature Gradient

1. Gradient of Total Head $\nabla \psi_t$

Liquid water flow in soils is formulated by

$$q_L = -K \nabla \psi_t \tag{6.48}$$

where $\nabla \psi_t$ is the gradient of total head ψ_t, which is defined by the sum of matric head ψ_m, osmotic head ψ_o, and gravitational head ψ_g as

$$\psi_t = \psi_m + \psi_o + \psi_g \tag{6.49}$$

resulting in

$$\nabla \psi_t = \nabla \psi_m + \nabla \psi_o + 1 \tag{6.50}$$

Assuming that both water content and solute concentration are constant in a soil, the gradients of matric head and osmotic head are given, respectively, by

$$\nabla \psi_m = \frac{\partial \psi_m}{\partial T} \nabla T \tag{6.51}$$

and

$$\nabla \psi_o = \frac{\partial \psi_o}{\partial T} \nabla T \tag{6.52}$$

The gradient of total head is then given by

$$\nabla \psi_t = \left(\frac{\partial \psi_m}{\partial T} + \frac{\partial \psi_o}{\partial T} \right) \nabla T + 1 \tag{6.53}$$

2. Gradient of Matric Head $\nabla\psi_m$

It is generally accepted that the matric potential of water in soils depends lineally on the surface tension of water σ_w (N m^{-1}). Hence, the first term of Equation (6.53) is given by

$$\frac{\partial \psi_m}{\partial T} = \frac{\psi_m}{\sigma_w}\frac{\partial \sigma_w}{\partial T} \tag{6.54}$$

Figure 6.13 shows surface tension of water as a function of temperature, which is approximated by a linear function

$$\sigma_w = (75.6 - 0.154T) \times 10^{-3} \tag{6.55}$$

Substituting Equation (6.55) into Equation (6.54), we obtain

$$\frac{\partial \psi_m}{\partial T} = -0.154 \times 10\frac{\psi_m}{\sigma_w} \tag{6.56}$$

The integral of Equation (6.56)

$$\int \frac{d\psi_m}{\psi_m} = -\int \frac{0.154}{75.6 - 0.154T}dT \tag{6.57}$$

gives the solution

$$\psi_m = C_m(75.6 - 0.154T) \tag{6.58}$$

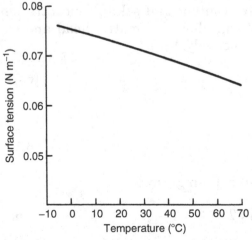

Figure 6.13 Surface tension of water as a function of temperature.

and its derivative

$$\frac{\partial \psi_m}{\partial T} = -0.154 C_m \qquad (6.59)$$

where C_m is the integral constant.

Kasubuchi and Miyazaki (12) collected experimental data indicating the change in matric head with temperature. These data are exhibited in Figure 6.14, where the theoretically predicted linear lines, given by Equation (6.58) for several values of C_m, are presented by thin solid lines together with experimentally obtained data presented by thick solid lines and dashed lines. The solid lines were obtained from the independently measured moisture characteristic curves of

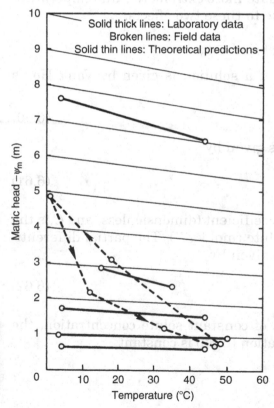

Figure 6.14 Change in matric head with temperature.

individual soils under different temperatures, and hence the volumetric water contents on each solid line were constant.

The dashed line, on the other hand, was obtained by using the same tensiometer–soil–water system in order to measure the change in matric head with a decrease in temperature from 48.5 to 1°C and a subsequent increase in temperature from 1 to 47°C (13). As much as 13 h was required to reach temperature equilibrium in the tensiometer–soil–water system. Gardner (13) attributed the marked change in matric head with temperature both to the direct effect of temperature on ψ_m, which is explained by Equation (6.58), and to the indirect effect of temperature, in which the thermally induced moisture movement has an influence on the change in ψ_m. It is generally known that matric head data obtained in fields fluctuate more extremely with temperature change, presumably due to these two effects.

3. Gradient of Osmotic Head $\nabla \psi_o$

The osmotic potential of a solution is given by Van't Hoff's equation,

$$\phi_o = -\nu CRT \tag{6.60}$$

and the osmotic head is given by

$$\psi_o = \frac{\phi_o}{g} \tag{6.61}$$

where ν is the osmotic coefficient (dimensionless) and C is the concentration of the solute (mol kg^{-1}). The partial differentiation of ψ_o by T is then given by

$$\frac{\partial \psi_o}{\partial T} = -\frac{\nu CR}{g} \tag{6.62}$$

Under the assumption of constant solute concentration, the right-hand side of Equation (6.62) is constant.

By substituting Equations (6.59) and (6.62) into Equation (6.53), we obtain

$$\nabla \psi_t = -\left(0.154C_m + \frac{\nu CR}{g}\right)\nabla T + 1 \tag{6.63}$$

Equation (6.63) is the driving force of liquid water flow in soil under the assumption that both water content and solute concentration are constant throughout the soil and that a temperature gradient exists.

B. Coefficients of Flow Under a Temperature Gradient

By substituting the rewritten total head gradient (Equation (6.63)) into Equation (6.48), the flow equation of liquid water under a temperature gradient is given by

$$q_{LT} = -D_{LT}\nabla T - K \tag{6.64}$$

where D_{LT} is the thermal liquid diffusivity (m^2 s^{-1} K^{-1}). It should, however, be taken into account that without a salt sieving effect, which restricts the movement of solute molecules within soil pores, no liquid water movement due to the gradient of solute concentration will be caused. When the restriction of solute movement is incomplete, part of the liquid water will migrate due to the solute concentration gradient. Under such conditions, the osmotic efficiency coefficient σ is given by (14)

$$\sigma = \frac{r_s - r_w}{b - r_w}, \quad 0 \le \sigma \le 1 \tag{6.65}$$

where r_s is the hydrated radius of solute, r_w is the radius of the water molecule, and b one-half of the water film thickness. To define the coefficients D_{LT} in Equation (6.64), the osmotic efficiency coefficient σ should be included appropriately as

$$D_{LT} = -K\left(0.154C_m + \frac{\sigma \nu CR}{g}\right) \tag{6.66}$$

C. Liquid Water Flow Due to Water Content Gradient, Solute Concentration Gradient, and Temperature Gradient

Generally, there are water content gradients, solute concentration gradients, and temperature gradients in fields, and liquid water moves due to these driving forces. Hence, the equation of liquid water flow in soils is generally given by

$$q_{\mathrm{L}} = -D_{\mathrm{LT}}\nabla T - D_{\mathrm{Lm}}\nabla\theta + D_{\mathrm{Lo}}\nabla C - K \qquad (6.67)$$

where D_{LT} is the thermal liquid diffusivity (m^2 s^{-1} K^{-1}), D_{Lm} is the isothermal liquid diffusivity (m^2 s^{-1}), and D_{Lo} is the liquid water diffusivity induced by the solute concentration gradient (m^2 kg s^{-1} mol^{-1}). As mentioned above, only when the soil pores have salt sieving effects does liquid water flow exist in proportion to the gradient of solute concentration. The salt sieving effect is expressed by means of an osmotic efficiency coefficient.

A reliable set of the coefficients of Equation (6.67) is given as

$$D_{\mathrm{LT}} = -K\left(0.154 C_{\mathrm{m}} + \frac{\sigma\nu CR}{g}\right) \qquad (6.68)$$

$$D_{\mathrm{Lm}} = K\frac{\partial\psi_{\mathrm{m}}}{\partial\theta} \qquad (6.69)$$

and

$$D_{\mathrm{Lo}} = \frac{K\sigma\nu RT}{g} \qquad (6.70)$$

Note that osmotic efficiency coefficient σ is included in both coefficients D_{LT} and D_{Lo}.

IV. OBSERVATIONS OF WATER FLOW IN SOILS UNDER TEMPERATURE GRADIENTS

A. Observations in Closed Soil Columns

The separation of liquid water flux q_{L} and vapor flux q_{V} in a laboratory may provide a good evaluation of the applicability of many proposed theories mentioned above. Closed soil columns with different temperatures at both ends have been widely chosen for this purpose.

Gurr et al. (6) applied an average temperature gradient of $1.6°C\ cm^{-1}$ to loam soil packed in a Perspex cylinder 10 cm in length and 14 cm in internal diameter. The initial water contents of the soil ranged from 1.7 to 24.5%. The chloride was used as a tracer of liquid water movement in all the runs. After 5 days of each run, they obtained data shown in Figure 6.15 where both water contents (black circles) and chlorides profiles (white circles) are shown. In this figure, they found that liquid water tends to accumulate at the low-temperature side of the column when the initial water contents are between 4.3 and 9.6%, while chlorides tend to accumulate at the high-temperature side of the column when initial water contents are between 7.9 and 19.8%. They concluded that the net transfer of water in a closed soil column under a temperature gradient is composed of water evaporation at the hotter soil, vapor movement into the colder soil, and condensation and return as a liquid at the colder soil.

They pointed out that the increase of water content at the low-temperature side of the soil column was 3.6 times more than the potential water vapor flux from the warm side to the cold side. It is practically suitable for us to follow and verify their calculation. In column C of Figure 6.15, a vapor flow of $0.18\ g\ cm^{-2}$ took place in 5 days across the plane at which the initial water content remained unchanged. Applying Equation (6.71) and Table 6.2, we obtain the effective diffusion coefficient D_{eff}, which is $2.65 \times 10^{-8}\ kg\ s^{-1}\ m^{-1}\ K^{-1}$ at an average temperature of 18°C. Other factors in Equation (6.71) are easily determined. The average air content a is 0.417 when the porosity of the soil is 0.47, ξ is 0.66 after Penman (3), and the temperature gradient is $-160°C\ m^{-1}$. The resultant water vapor flux is $1.167 \times 10^{-6}\ kg\ s^{-1}\ m^{-2}$ and the total amount of water flow during 5 days may be $0.05\ g\ cm^{-2}$, which is only 28% of the estimated vapor flow.

Gurr et al. (6) further described that the estimated vapor flow of $0.18\ g\ cm^{-2}$ is still an underestimate of the vapor flow because of reverse liquid flow.

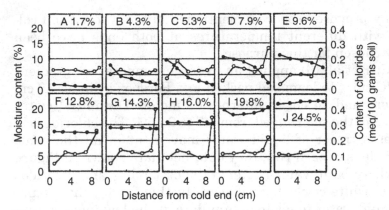

Figure 6.15 Distribution of water and chlorides in columns of loam soil, of varying initial water contents, subjected to a temperature gradient for 5 days. (After Gurr, C. G., Marshall, T. J., and Hutton, J. T., *Soil Sci.* 74:335–345 (1952). With permission.)

Observations in closed soil columns by Gurr et al. (6), Taylor and Cavazza (7), Jackson et al. (15), and many other researchers revealed the excess vapor flows than predictions (Equation (6.71)) under given temperature gradients.

B. Observations in Open Soil Columns

In addition to the separation of liquid water flux q_L and vapor flux q_V, the accurate measurement of reverse liquid flow from a clod soil provides more precise estimation of vapor flow under a temperature gradient. Alternatively, the direct measurement of vapor flux q_V in a soil under a temperature gradient is more suitable for the selection of a reliable equation of water vapor movement under a temperature gradient.

Miyazaki (16) applied a one-dimensional temperature gradient, $1.35 \pm 0.10°C$ cm^{-1}, along the two columns of 10 cm height and 10 cm diameter, whose surface was open and side walls were adequately insulated, in a small room where the temperature and relative humidity were kept at 37°C and $87.5 \pm 2.5\%$, respectively, as illustrated in Figure 6.16. The bottoms of the two columns were kept at lower temperature, to generate water vapor condensations there, by applying water of 20°C through flexible tubes. One of the columns

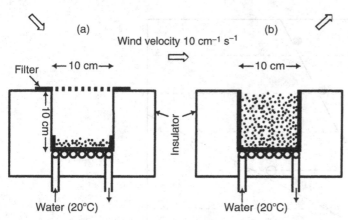

Figure 6.16 Distribution of water and chlorides in columns of loam soil, of varying initial water contents, subjected to a temperature gradient for 5 days. (After Gurr, C. G., Marshall, T. J., and Hutton, J. T., *Soil Sci.* 74:335–345 (1952). With permission.)

was filled with 1-cm thick air-dry sand and the other with 10-cm thick air-dry sand. The cumulative condensations of water vapor into both sand columns must be equal to the cumulative water vapor fluxes at the surface zones of each column. These fluxes were calculated by applying Equation (6.29), identical to Equation (6.71), at the surface zones of both columns. The air content of the surface zones of the left and right columns were 1.00 and 0.45, and the tortuosities ξ were 1.00 and 0.66, respectively, and the relative humidities inside both columns were assumed to be 1.00. Figure 6.17 shows the measured cumulative condensations of water vapor into both sands and the predicted cumulative water vapor fluxes. The excellent agreements suggest that Equation (6.29) is applicable to the water vapor flow both in a static air space and in air-dry sand under temperature gradients.

Figure 6.18 shows the changes in measured and predicted water contents in the bottom 1-cm layer of the 10-cm thick sand column whose initial water content was 1.2%. The prediction by Equation (6.29), with the air content being 0.43, the tortuosity ξ being 0.66, and the relative humidity being 1.00, underestimated the measured water content especially

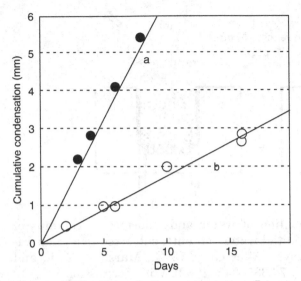

Figure 6.17 Cumulative water vapor fluxes q_{VT} in columns a and b (solid lines), and corresponding cumulative condensations of water vapor to each column (dots). (After Miyazaki, T., *Trans. Jpn. Soc. Irrig. Drain. Reclam. Eng.* *61*:1–8 (1976). With permission.)

at the initial few days. The prediction with the modified model by Philip and de Vries (10), given by Equations (6.33) to (6.35), was better in the early stage of condensation but overestimated the measured water content in the latter stage presumably due to the reverse liquid flow associated with the condensation. Taking account of a little reverse liquid flow, recognized even after 1 day from the start of the measurement, Miyazaki (16) recommended the use of the modified models by Philip and de Vries (10) when the sand is moderately wet.

C. Observations in a Field

The distinction of water vapor flux from liquid flux in a field under temperature gradients is a matter of practical concern especially in arid and semiarid areas. To exemplify this distinction, we can reasonably estimate these fluxes in a sand

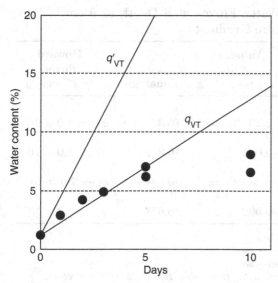

Figure 6.18 Change in water content at the bottom 1-cm sand layer of column b whose initial water content is 1.2%. Equations (6.29) and (6.33) implemented the predictions of q_{VT} and q'_{VT}, respectively. (After Miyazaki, T., *Trans. Jpn. Soc. Irrig. Drain. Reclam. Eng.* *61*:1–8 (1976). With permission.)

dune field under the conditions given in Figure 6.1 and Figure 6.2, based on the discussion presented in this chapter.

To tackle the question more concretely, we estimate both water vapor fluxes and liquid water fluxes across the depth of 9 cm of the sand dune of Figure 6.2 during 1:00 a.m. and 3:00 a.m., when temperature increases with depth, and soil moisture increases from 0 to 9 cm and decreases from 9 to 25 cm. Figure 6.19 gives the illustrative moisture profiles changing between 1:00 and 3:00 a.m. The average temperature at 9-cm depth is 30.5°C, the average temperature gradient is -0.23°C cm^{-1}, and the average volumetric water content is 0.068 cm^3 cm^{-3} during this time interval.

Table 6.1 shows the results both for vapor phase fluxes and liquid phase fluxes. The rate of water vapor condensation, Q, from the atmosphere to the land surface at 2:00 a.m., when the wind velocity was zero, the relative humidity was 0.96, the temperature was 23.6°C, and the temperature gradient was

Table 6.1 Estimated Water Fluxes at a Depth of 9 cm of Sand Dune Under a Temperature Gradient

Phases of water	Driving forces	Values of driving forces	Equations	Upward fluxes $(g\ s^{-1}\ cm^{-2})$
Vapor	Temperature gradient	-0.23 (°C cm^{-1})	(6.33)[a]	$q'_{VT} = 1.0 \times 10^{-7}$
Vapor	Moisture gradient	-0.002	(6.30)	$q_{Vm} = 1.0 \times 10^{-13}$
Liquid	Temperature gradient	-0.23 (°C cm^{-1})	(6.64)[b]	$q_{LT} = 7.0 \times 10^{-7}$
Liquid	Moisture gradient	-0.002	(6.67)[c]	$q_{Lm} = 7.3 \times 10^{-6}$

[a]Philip and de Vries model is used.
[b]Osmotic efficiency coefficient σ is zero.
[c]D_{LT} and D_{LO} are zero. K is negligible.
Source: Miyazaki, T. and Amemiya, Y., *Trans. Jpn. Soc. Irrig. Drain. Reclam. Eng.48*:16–22 (1973). With permission.

$0.02°C\ cm^{-1}$ between 5 and 35 cm above the land surface, was estimated to be $4.8 \times 10^{-9}\ g\ s^{-1}\ cm^{-2}$. The degree of contribution to the increase of water content between 0 and 9 cm of

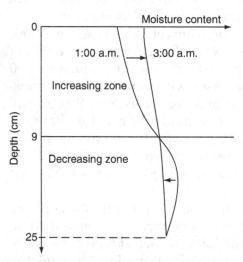

Figure 6.19 Illustrative moisture profiles at 1:00 and 3:00 a.m. in a sand dune.

the sand dune during 1:00 and 3:00 a.m. was thus estimated (1) to be

$$q_{Lm} > q_{LT} > q_{VT} > Q > q_{Vm}$$

D. An Observation with Salt Effects

Figure 6.20 is an example of experimental data presented by Nassar and Horton (17), who maintained the hot and cold ends of a horizontal column packed with Ida silt loam at 19.02 and 8.93°C, respectively. Water flow in this closed system under a temperature gradient is shown schematically in Figure 6.20. The liquid flux q_{LT} and water vapor flux q_{VT}, both induced by a temperature gradient, will proceed to the cold end, while other fluxes — q_{Vm}, water vapor flux induced by matric potential gradient; q_{Lm}, liquid water flux induced by matric potential gradient; q_{Vo}, water vapor flux induced by osmotic potential gradient; and q_{Lo}, liquid water flux induced by osmotic potential gradient — will all proceed to the hot end. In the hot zone of the column, evaporation of water will promote the accumulation of salt. In the cold zone of the column, condensation of water vapor will promote the accumulation of water. After sufficient time the soil column will arrive at a steady state in which all the fluxes are balanced.

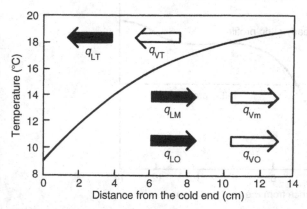

Figure 6.20 Profile of temperature and corresponding liquid and vapor flows. (From Nassar, I. N. and Horton, R., *Soil Sci. Soc. Am. J.* *53*:1323–1329 (1989). With permission.)

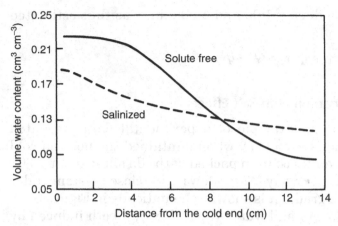

Figure 6.21 Moisture profiles of solute-free water and salinized water. (From Nassar, I. N. and Horton, R., *Soil Sci. Soc. Am. J.* *53*:1323–1329 (1989). With permission.)

Figure 6.21 shows the moisture profiles of solute-free water and salinized water in the steady state. The initial water content was 0.144 m^3 m^{-3} in both cases. The profile of solute-free water is more biased than the profile of salinized water. This biasing is presumably attributed to the accumulation of solute in the hot zone of the column shown in Figure 6.22, which causes the increases of q_{Lo} and q_{Vo}.

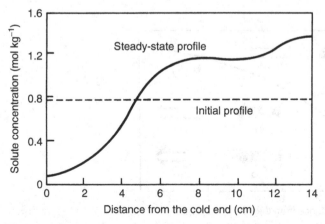

Figure 6.22 Initial and steady solute concentration profiles. (From Nassar, I. N. and Horton, R., *Soil Sci. Soc. Am. J.* *53*:1323–1329 (1989). With permission.)

Furthermore, Nassar and Horton (17) calculated the value of the osmotic efficiency coefficient σ as a function of water film thickness by using Equation (6.65). The values of σ ranged from 0.14 at the cold end of the column to 0.28 at the hot end of the column. Precise measurements of temperatures, water contents, and solute concentrations in soils promote the elucidation of transport phenomena in unsaturated soils under temperature gradients.

APPENDIX: PRACTICAL ESTIMATION OF VAPOR FLUX

It is often necessary to make a quantitative estimation of water vapor flux in soils, at least an estimation of its order, under a temperature gradient, without knowing the values of many physical parameters contained in the equations of flow. Nevertheless, since some of the parameters in these equations are functions of temperature and pressure, even the simple equation of thermally induced water vapor flow given by Equation (6.29) is not necessarily simple enough to obtain an approximate value of the flux.

The following procedure is therefore convenient for estimating the approximate water vapor flux under temperature gradients. When water vapor is saturated in the pores of the soil, Equation (6.29) is written as

$$q_{\mathrm{VT}} = -D_{\mathrm{eff}} a \xi \frac{\mathrm{d}T}{\mathrm{d}z} \tag{6.71}$$

where a is the volumetric air content, ξ is the tortuosity (which is 0.66 [after Penman]), and D_{eff} is the effective diffusion coefficient, defined by

$$D_{\mathrm{eff}} = \frac{\gamma D p_0 H}{R_v{}^2 T^3} \tag{6.72}$$

Table 6.2 gives the values of D_{eff} (kg s^{-1} m^{-1} K^{-1}) together with the values of parameters contained in Equation (6.72) as functions of temperature. All the values are given in SI units. The unit of D_{eff} is transformed into m^2 s^{-1} K^{-1} by dividing

Table 6.2 Values of D_{eff} Defined by Equation (6.72) as a Function of Temperature

Temperature, T (°C)	Diffusion coefficient of water vapor, D ($m^2\,s^{-1}$)	Saturated water vapor pressure, p_0 (Pa)	Latent heat of evaporation, H ($J\,kg^{-1}$)	Mass flow factor, γ	Effective diffusion coefficient, D_{eff} ($kg\,s^{-1}\,m^{-1}\,K^{-1}$)
0	0.234×10^{-4}	611	2.500×10^6	1.006	8.28×10^{-9}
10	0.253×10^{-4}	1227	2.477×10^6	1.012	1.61×10^{-8}
11	0.256×10^{-4}	1312	2.475×10^6	1.013	1.72×10^{-8}
12	0.258×10^{-4}	1402	2.472×10^6	1.014	1.83×10^{-8}
13	0.260×10^{-4}	1497	2.470×10^6	1.015	1.96×10^{-8}
14	0.262×10^{-4}	1598	2.468×10^6	1.016	2.08×10^{-8}
15	0.264×10^{-4}	1705	2.465×10^6	1.017	2.21×10^{-8}
16	0.266×10^{-4}	1818	2.463×10^6	1.018	2.36×10^{-8}
17	0.268×10^{-4}	1937	2.461×10^6	1.019	2.51×10^{-8}
18	0.270×10^{-4}	2064	2.458×10^6	1.021	2.65×10^{-8}
19	0.272×10^{-4}	2197	2.456×10^6	1.022	2.83×10^{-8}
20	0.274×10^{-4}	2338	2.453×10^6	1.024	3.00×10^{-8}
21	0.277×10^{-4}	2487	2.451×10^6	1.025	3.20×10^{-8}
22	0.279×10^{-4}	2644	2.449×10^6	1.027	3.38×10^{-8}
23	0.281×10^{-4}	2810	2.446×10^6	1.029	3.59×10^{-8}
24	0.283×10^{-4}	2984	2.444×10^6	1.030	3.80×10^{-8}
25	0.285×10^{-4}	3168	2.442×10^6	1.032	4.03×10^{-8}
26	0.288×10^{-4}	3362	2.440×10^6	1.034	4.28×10^{-8}
27	0.290×10^{-4}	3566	2.437×10^6	1.036	4.54×10^{-8}
28	0.292×10^{-4}	3781	2.435×10^6	1.039	4.80×10^{-8}
29	0.294×10^{-4}	4007	2.433×10^6	1.041	5.07×10^{-8}
30	0.296×10^{-4}	4045	2.430×10^6	1.044	5.36×10^{-8}
31	0.299×10^{-4}	4495	2.428×10^6	1.046	5.69×10^{-8}
32	0.301×10^{-4}	4757	2.425×10^6	1.049	6.02×10^{-8}
33	0.303×10^{-4}	5033	2.423×10^6	1.052	6.36×10^{-8}
34	0.306×10^{-4}	5322	2.421×10^6	1.055	6.74×10^{-8}
35	0.308×10^{-4}	5626	2.418×10^6	1.059	7.11×10^{-8}
36	0.310×10^{-4}	5945	2.416×10^6	1.062	7.51×10^{-8}
37	0.312×10^{-4}	6279	2.414×10^6	1.066	7.93×10^{-8}
38	0.315×10^{-4}	6630	2.412×10^6	1.070	8.40×10^{-8}
39	0.317×10^{-4}	6997	2.409×10^6	1.074	8.85×10^{-8}
40	0.319×10^{-4}	7381	2.407×10^6	1.079	9.35×10^{-8}

each value by the density of water (1 Mg m^{-3}). In the calculation, the gas constant of water vapor R_v was set at 461.9 J $\text{kg}^{-1} \text{K}^{-1}$.

Table 6.2 may be useful in estimating the approximate value of water vapor flux in the soils. For example, if the temperature gradient of a given soil whose air content is 0.5 $\text{m}^3 \text{m}^{-3}$ is 100 K m^{-1} ($1°C \text{ cm}^{-1}$ in traditional units), the flux of water vapor at 30°C is estimated by

$$q_{VT} = -5.36 \times 10^{-8} \times 0.5 \times 0.66 \times (-100)$$
$$= 1.77 \times 10^{-6} (\text{kg s}^{-1} \text{ m}^{-2})$$
$$= 1.77 \times 10^{-7} (\text{g s}^{-1} \text{ cm}^{-2})$$

Thus, the rough estimation of thermally induced water vapor flux is obtained by giving the value of D_{eff} in Table 6.2 the air content, and the temperature gradient of given soils. This estimation is applicable to such air-dry soils as sand in deserts and sand dunes, but underestimates the fluxes in moderately wet soils.

REFERENCES

1. Miyazaki, T. and Amemiya, Y., Effects of temperature and condensation on the soil moisture movement, *Trans. Jpn. Soc. Irrig. Drain. Reclam. Eng.* 48:16–22 (1973).
2. Brawand, H. and Kohnke, H., Microclimate and water vapour exchange at the soil surface, *Soil Sci. Soc. Am. Proc.* 16:195–198 (1952).
3. Penman, H. L., Gas and vapour movements in the soil. 1. The diffusion of vapours through porous solids, *J. Agric. Sci.* 30:437–462 (1940).
4. Rollins, R. L., Spangler, M. G., and Kirkham, D., Movement of soil moisture under a thermal gradient, *Proc. Highway Res. Board* 33:492–508 (1954).
5. Nakano, M. and Miyazaki, T., The diffusion and nonequilibrium thermodynamic equations of water vapor in soils under temperature gradients, *Soil Sci.* 128(3):184–188 (1979).

6. Gurr, C. G., Marshall, T. J., and Hutton, J.T., Movement of water in soil due to a temperature gradient, *Soil Sci.* 74:335–345 (1952).

7. Taylor, S. A. and Cavazza, L., The movement of soil moisture in response to temperature gradients, *Soil Sci. Soc. Am. Proc.* 18:351–358 (1954).

8. Cass, A., Campbell, G. S., and Jones, T. L., Enhancement of thermal water vapor diffusion in soil, *Soil Sci. Soc. Am. J.* 48:25–32 (1984).

9. Iwata, S., Tabuchi, T., and Warkentin, B. P., *Soil–Water Interactions,* Marcel Dekker, New York (1988).

10. Philip, J.R. and de Vries, D. A., Moisture movement in porous materials under temperature gradients, *Trans. Am. Geophys. Union* 38:222–228 (1957).

11. July, W. A. and Letey, J., Water vapor movement in soil: reconciliation of theory and experiment, *Soil Sci. Soc. Am. J.*43:823–827 (1979).

12. Kasubuchi, T. and Miyazaki, T., Temperature and thermal transfer, in *Stuchi-no-Butsurigaku,* Research Association of Soil Physics, Japan, Ed., Morikita-Shuppan Ltd., Tokyo, pp. 279–293 (1979).

13. Gardner, R., Relation of temperature of moisture tension of soil, *Soil Sci.* 79:257–265 (1955).

14. Nassar, I. N. and Horton, R., Water transport in unsaturated nonisothermal salty soil. II. Theoretical development, *Soil Sci. Soc. Am. J.* 53:1330–1337 (1989).

15. Jackson, R.D., Rose, D.A., and Penman, H.L., Circulation of water in soil under a temperature gradient, *Nature* 205:314–316 (1965).

16. Miyazaki, T., Condensation and movement of water vapor in sand under temperature gradient, *Trans. Jpn. Soc. Irrig. Drain. Reclam. Eng.* 61:1–8 (1976).

17. Nassar, I. N. and Horton, R., Water transport in unsaturated nonisothermal salty soil. I. Experimental results, *Soil Sci. Soc. Am. J.* 53:1323–1329 (1989).

7

Effects of Microbiological Factors on Water Flow in Soils

TSUYOSHI MIYAZAKI, KATSUTOSHI SEKI

Department of Biological and Environmental Engineering,
The University of Tokyo, Japan

I. TYPICAL CHANGE IN PERMEABILITY

It is generally known that when a soil is submerged and the hydraulic conductivity of the soil is measured every day, a systematic change of saturated hydraulic conductivity is obtained, as shown in Figure 7.1. Allison (1) explained this change in permeability as consisting of the following three phases:

1. *Initial decrease attributed to a structural change.* The decrease in this permeability is probably due either to structural changes resulting from swelling and dispersion of the dry soil upon wetting or due to dispersion resulting from a decrease in electrolyte

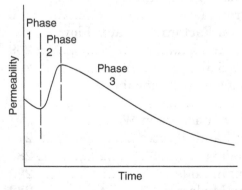

Figure 7.1 Systematic daily change in the saturated hydraulic conductivity of submerged soil. (After Allison, L. E., *Soil Sci. 63*:439–450 (1947). With permission.)

content of the soil solution as any salts present are removed in the percolate. On highly permeable soils this initial decrease is small, or nonexistent, but for relatively impermeable soils, it may be appreciable and continue for 10 to 20 days before the second phase of increase is apparent.

2. *Increase attributed to the removal of entrapped air.* When soils are submerged, considerable air is trapped in the pores. As the air is dissolved and removed in the percolating water, the permeability gradually increases, attaining a maximum when all or nearly all of the entrapped air is removed. The minimum permeability at the turning point from phase 1 to phase 2 appears to be due to two opposing phenomena; that is, the forces in phase 1 tending to reduce permeability from the beginning and the forces in phase 2 tending to increase permeability.

3. *Gradual decrease of permeability following phase 2.* After the maximum is reached, the permeability decreases with time, rather rapidly at first, then more slowly. According to Allison (1), one or more of the following three contributing causes are involved in this decrease:

1. Slow physical disintegration of aggregates under prolonged submergence.

2. Biological clogging of soil pores with microbial cells and their synthesized products, slimes, or polysaccharides.

3. A dispersion due to the attack of microorganisms on organic materials that bind soil into aggregates.

II. MICROBIOLOGICAL EFFECTS ON PERMEABILITY

A. Direct Effects

There are two types of microbiological effects on the permeability of soils: direct and indirect. The direct biological factors that influence the decrease in hydraulic conductivity are the production of gases, the microbial destruction of soil structure, and the accumulation of metabolic products in soil pores.

Martin (2) attributed the destruction of soil structure to the attack by microorganisms on organic materials, resulting in a dispersion of soil particles. He also found that the production of metabolic products was enhanced by the addition of an energy source such as sucrose. Hydraulic conductivity of soil treated with such an energy source decreased at a rapid rate compared with untreated soil. Poulovassilis (3) reported that the gases produced by microorganisms have a major influence on the decrease in permeability.

Frankenberger et al. (4) investigated the relation between bacterial numbers and hydraulic conductivities. Table 7.1, extracted from their data, shows relations between anaerobic bacterial numbers per gram of oven-dried soils sampled from the reduced zone of the soil column and the entire hydraulic conductivities of corresponding soils. Bacterial numbers were determined by dilution plate counts on egg-albumin agar under anaerobic conditions. To sterilize soil columns, the soils were placed in an autoclave and heated for 30 min at 121°C. To fertilize soil columns, they were leached with 500 ppm glucose continuously over 2000 h. Soil columns leached with distilled water only were used for comparison.

Table 7.1 Relations Between Anaerobic Bacterial Numbers per Gram of Oven-Dried Soils and Hydraulic Conductivities of Two Soils

Continuous submergence with:	Nicollet loam		Tama silty clay loam	
	Anaerobic count (g^{-1})	Saturated hydraulic conductivity $(cm\ s^{-1})$	Anaerobic count (g^{-1})	Saturated hydraulic conductivity $(cm\ s^{-1})$
Sterile water	0	2.64×10^{-2}	0	2.25×10^{-2}
Distilled water	11×10^4	1.12×10^{-3}	7×10^4	7.36×10^{-4}
Glucose-water	11×10^5	2.22×10^{-5}	22×10^4	3.89×10^{-5}

Source: After Frankenberger, W. T., Jr., Troeh, F., and Dumenil, L. C., *Soil Sci. Soc. Am. J. 43*:333–338 (1979). With permission.

It is evident from Table 7.1 that leaching with glucose caused a marked increase in anaerobic bacteria, which caused a decrease in the hydraulic conductivities of both soils. Furthermore, they demonstrated that the hydraulic conductivity of Tama silty clay loam decreased sharply with time as shown in Figure 7.2 when nutrients and energy sources such as KNO_3 and glucose were employed. Because of the initial treatment in this experiment, removal of the trapped air (phase 2 in Figure 7.1) did not cause an early increase in hydraulic conductivities.

B. Indirect Effects

1. Effects of Gaseous Production

Continuous submergence of water makes the soil environment anaerobic, and this anaerobic condition potentially

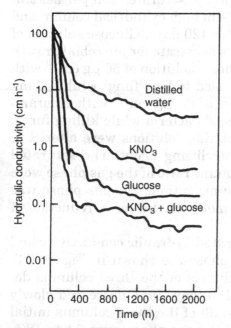

Figure 7.2 Effect of various treatments on the hydraulic conductivities of Tama silty clay loam under a continuous treatment. (After Frankenberger, W. T., Jr., Troeh, F., and Dumenil, L.C., *Soil Sci. Soc. Am. J. 43*:333–338 (1979). With permission.)

produces methane gas due to microbial anaerobic respiration especially when the soil is rich in carbon content. Methane production is typically seen in subsurface peat soil of wetland, and it is plausible to assume that the methane gas is partly stored as gas bubbles. Reynolds et al. (5) found that, in their experiments with three columns and five runs under continuous water flow over time periods ranging from 44 to 78 days, the saturated hydraulic conductivity of peat decreased on average by 1.6 orders of magnitude, and at the same time volumetric water content decreased by about 20 percentage points, and gaseous methane concentration increased from 4 to 50 μmol ml^{-1}.

Another typical anaerobic condition under continuous submergence is also observed in paddy rice field, where methane gas is likely produced. Seki et al. (6) conducted a series of column experiments with paddy field soil (Andisol) to observe the biological clogging on the saturated soil permeability. They packed the soil in a 1-cm high cylindrical column and percolated nutrient solution for 120 days. Glucose solution of 50 μg cm^{-3} (GCO) was used as substrate for microbial growth of both bacteria and fungi, glucose solution of 50 μg cm^{-3} with chloramphenicol (GCH) was used to feed fungi while killing bacteria, and glucose solution of 50 μg cm^{-3} with chloramphenicol (GCY) was used to feed bacteria while killing fungi. After 120 days, all the percolating solutions were altered to sodium azide solution as a sterilizing agent. The saturated hydraulic conductivity and volume ratio of the gas phase were measured every day. The volume ratio of the gas phase was calculated from the measurement of total mass reduction of the column.

The changes in the saturated hydraulic conductivity and the volume ratio of the gas phase are shown in Figure 7.3. Saturated hydraulic conductivities of the three columns decreased rapidly in the initial 10 days, and decreased slowly in the remaining 110 days. In all of the three columns initial values of the volume ratio of the gas phase were 14 to 18%, and they decreased by dissolving into the percolating water at the earliest stage of percolations. In the GCO run, where only glucose solution was supplied, the volume ratio of the gas

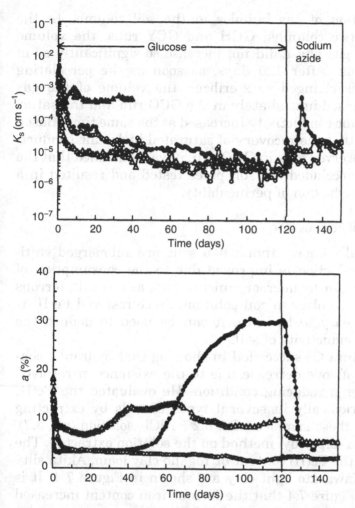

Figure 7.3 Changes in saturated hydraulic conductivity K_s and volume ratio of the gas phase a. Percolating solutions are GCO (•), GCH (○), and GCY (△). At 120 days, the sterilizer was supplied. Symbols (▲) are from reference data. (After Seki, K., Miyazaki, T., and Nakano, M., *Eur. J. Soil Sci.* 49:231–236 (1998). With permission.)

phase increased from 3.6% at 40 days to 30.6% at 103 days. After 103 days the volume ratio of the gas phase seemed to have been stabilized. Therefore, the hydraulic conductivity decrease in the GCO run after 40 days may be attributed to the production of gas bubbles in the soil column. In the remaining two columns, GCH and GCY runs, the volume ratio of the gas phase did not increase so significantly as in the GCO run. After 120 days, as soon as the percolating solution was changed to sterilizer, the volume of the gas phase decreased immediately in the GCO run and the saturated hydraulic conductivity increased at the same time. Based on the fact that the recovery of saturated hydraulic conductivity was observed only in GCO run, it was concluded that the gas bubbles occluded the soil pores tested and resulted in a significant reduction of permeability.

2. Effects of Ferrous Iron

It is generally known that when soils are submerged statically, soil reduction is improved due to the consumption of oxygen in water by microorganisms, and, as a result, ferrous iron [Fe(II)] resolves in soil solution. Since resolved Fe(II) is active and easy to hydrate, it can be used to deduce the hydraulic conductivity of soil.

Motomura (7) succeeded in showing that hydraulic conductivities of soils decrease due to the existence of resolved Fe(II) under a reducing condition. He evaluated the Fe(II) content periodically in several types of soils by extracting Fe(II) from these samples with 0.2% $AlCl_3$ solution (pH 3.7), using an α,α'-dipyridyl method on the solution extracted. The changes in the Fe(II) content of Togane clay loam, Akita silty clay, and Kawazato light clay are shown in Figure 7.4. It is clear from Figure 7.4 that the ferrous iron content increased with time in soils submerged but that the percentages increase differed among soils. These differences were attributed by Motomura (7) to the difference in soil reduction.

Table 7.2 shows changes in the ferrous iron, Fe(II), contents in run 1 where disturbed Kawazato light clay was packed in a test bottle, stirred well, and submerged for 38 days under 30°C. While the Fe(II) content increased in the soil, the pH value of the skimmed water slightly increased and its Eh value remarkably decreased during this run. In run 2, disturbed Kawazato light clay was also submerged for 40 days

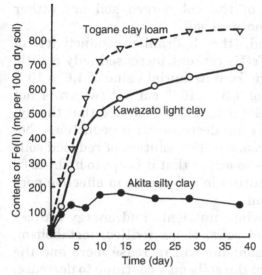

Figure 7.4 Change of contents of ferrous iron [Fe(II)] in three soils. (From Motomura, S., *Bull. Natl. Inst. Agric. Sci. Ser. B* 21:1–114 (1969). With permission.)

Table 7.2 Changes in Physical and Chemical Properties of Kawazono Clay During Long-Term Submergences

	Item	Initial value	Final value
	Fe(II) (mg g^{-1} dry soil)	0.017	6.93
Run 1	In skimmed water		
	PH	5.4	6.7
	Eh (mV)	+455	−127

	Item	Natural field soil	Submerged packed soil
	Porosity (cm^3 cm^{-3})	0.544	0.627
Run 2	Saturated hydraulic		
	Conductivity (cm s^{-1})	4.9×10^{-4}	2.0×10^{-5}

Source: Motomura, S., *Bull. Natl. Inst. Agric. Sci. Ser. B* 21:1–114 (1969). With permission.

under 30°C and sampled without disturbance to know the effect of reduction on the saturated hydraulic conductivity. It is apparently a contradiction that the porosity of submerged soil is larger than that of the natural soil while the saturated hydraulic conductivities of the submerged soil are rather smaller than that of the natural soil.

On the other hand, the hydraulic conductivity of Akita silty clay, whose Fe(II) content increased only a little in 40 days, did not change from its initial value of 1.6×10^{-6} cm^{-1} to the final value of 1.3×10^{-6} cm s^{-1} (which is not indicated in Table 7.2). Motomura (7) concluded that the hydraulic conductivities of soils decrease with reductions, because Fe(II) tends to increase in the solution of reduced soils and the dissolved Fe(II) is so active that it is apt to hydrate or absorb to soil colloid, resulting in a decrease in effective pore space for water percolation.

It is expected that when nutrients and energy sources are added to the reduced soils, the activities, metabolism, and numbers of microorganisms will increase more and the hydraulic conductivities of the soils may continue to decrease. Motomura (7), Frankenberger et al. (4), and many other researchers confirmed this prediction by adding glucose, saccharose, or potassium nitrate to the percolating water and by recognizing the resulting decrease in hydraulic conductivities of soils with time. For example, the hydraulic conductivity of the reduced Kawazato light clay mentioned above continues to decrease, from 2.0×10^{-5} to 4.0×10^{-6} cm s^{-1}, on the addition of glucose.

The principal indirect effect on hydraulic conductivities of microbes in soils is thus that of soil reduction. Although this indirect effect is of major importance in rice paddies, which are so predominant in Asian agriculture, less attention has been paid to this effect. Mitsuchi (8) reported the vertical distribution of iron (free iron oxide) in volcanic ash soil samples taken from rice paddies and adjacent upland fields. Figure 7.5 shows an example of the distribution of iron plotted from their data. In upland fields, iron is distributed relatively uniformly, whereas there were sharp peaks in rice paddies at depths of 20 to 30 cm. This is known to be evidence of eluviation of iron from surface soil and its accumulation within subsoil, where traffic pans exist. There is neither eluviation nor accumulation of iron in upland fields, since ponded water is not required there, and therefore reductive subsoil will not

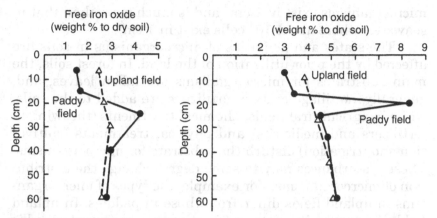

Figure 7.5 Profiles of iron contents in soils in paddies and adjacent upland fields. (From Mitsuchi, M., *Jpn. J. Soil Sci. Plant Nutr.* *41*(8):307–313 (1970). With permission.)

be generated. In forests, the iron distribution may be as uniform as in upland fields. Such an uneven distribution of iron in rice paddies will have an influence on the percolation of water, but little is known on this point.

III. EFFECTS OF MICROBIOLOGICAL FACTORS ON WATER FLOW IN NATURAL SOILS

A. Microorganisms in Soils

There are many kinds of microorganisms in soils. Among them, bacteria, fungi, and actinomycetes are predominant. There are, for example, about 10^7 to 10^8 bacteria in 1 g of dry soil, which decompose both organic and inorganic matter in soils, keeping a delicate ecological balance. Since the specific surfaces of soils range from about 10 to 100 m^2 g^{-1}, there may exist, by a simple calculation, about 10 to 1000 bacteria per 1 cm^2 of soil particle surface. The number of microorganisms is generally greatest at the land surface and decrease with soil depth to about one tenth of that value at a depth of 50 cm and to less than 1/1000 at a depth of 100 cm. It is evident that the number of microorganisms in soils is generally more than that in fertile lakes, where about 10^6 cells of microorganisms exist in 1 cm^3 and is much more than that in seawater, where about 10^3 cells exist in 1 cm^3.

The states and activities of microorganisms in soils are affected by the prior utilization of the land. In forest soils, the main substrates for microorganisms are fallen leaves, and, generally, no tillage and no fertilizers are added to the soils, while in agricultural fields, chemical treatments (throwing in fertilizers and medicines) and physical treatments (cultivation and irrigation) disturb the substrate for microorganisms. These disturbances may to some degree change the distribution of microorganisms. For example, the types of microorganisms in upland fields differ from those in paddies. In upland fields, as oxygen in soil pores is relatively abundant, aerobic bacteria are predominant, whereas in paddies, soils under ponded water are reduced and anaerobic bacteria are predominant.

Ishizawa and Toyoda (9) counted the number of bacteria in volcanic ash soil samples taken at every depth from forests, paddies, and upland fields in the Kanto area of Japan. Figure 7.6 is an example of the vertical distribution of bacteria counts. It is evident that the number of bacteria is highest at the land surface and decreases with depth exponentially. They noted that the counts of microorganisms were higher in arable

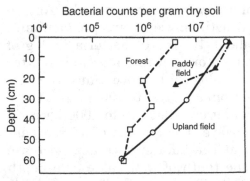

Figure 7.6 Vertical distributions of bacteria in fields of volcanic ash soils. (From Ishizawa, S. and Toyoda, H., *Bull. Natl. Inst. Agric. Sci. Ser. B 14*:203–284 (1964). With permission.)

soils (i.e., in paddies and upland fields) than in virgin soils (i.e., in forests).

Fujikawa et al. (10) counted the numbers of bacteria and fungi separately in a field under rotating use for rice and vegetable productions in Kyusyu area of Japan. Figure 7.7 shows their periodical vertical distributions. Special attention must be paid to the cropping schedule of this field where the soil had been submerged for a paddy use from May to September

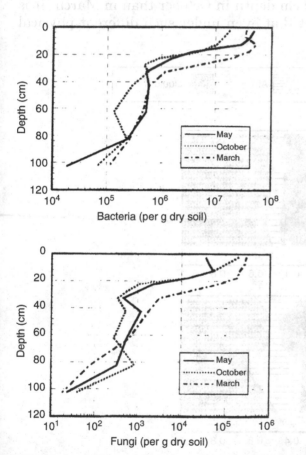

Figure 7.7 Seasonal change in count profiles of bacteria and fungi in a field under rotating use. (After Fujikawa, T., Miyazaki, T., and Imoto, H., *Trans. Jpn. Soc. Irrig. Drain. Reclam. Eng.* 71(3):111–118 (2003). With permission.)

and drained for an upland use from October to April every year. Figure 7.8 shows the three phase distributions of soils in both October when the air phase was very small due to the submergence and subsequent swelling of the soil under a paddy use, and in March when the air phase was relatively high due to the drainage and tillage after the harvest of rice. In addition, Fujikawa et al. (10) found that CO_2 concentration underneath was greatly higher in October than in March probably due to the higher average soil temperature from land surface to 100 cm depth in October than in March. It is a matter of interest that even under such different physical

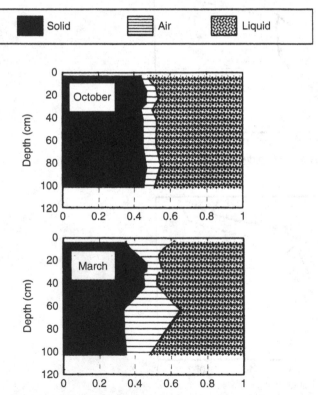

Figure 7.8 Seasonal change in solid, air, and liquid phase profiles in a field under rotating use. (After Fujikawa, T., Miyazaki, T., and Imoto, H., *Trans. Jpn. Soc. Irrig. Drain. Reclam. Eng.* 71(3):111–118 (2003). With permission.)

conditions the distributions of microbial counts are rather stable as are seen in Figure 7.7.

B. Microbiological Effects on Permeability of Natural Soils

One of the most popular techniques for investigating microbiological effects on the flow of water in soils is comparison of the difference in hydraulic conductivities between sterile soils and nonsterile soils. Soils submerged with water, dissolving sterilizers, or nutrients are also used for comparison of the difference in hydraulic conductivities. In sterile soils (or in soils treated with a water-soluble sterilizer), the hydraulic conductivities change with time due to physical and chemical factors, whereas in nonsterile soils (or in soils treated with water-soluble nutrients), the hydraulic conductivities change due to physical, chemical, and microbiological factors.

Miyazaki et al. (11) reported microbiological effects such as those noted above on soil permeabilities. They took four undisturbed volcanic ash soil samples at 10-cm increments from the land surface down to a depth of 60 cm in a forest, a rice paddy, and an upland field located close together in the Kanto area in Japan. Figure 7.9 shows profiles of hardness, water content, and bulk density in these fields. A sharp peak in hardness assumed to be the traffic pan appeared at a depth of 25 cm in the rice paddy. A similar but weak peak in hardness appeared at a depth of 20 cm in the upland field. The bulk density of the surface soil in the forest was lower than that of the subsurface soil, whereas those in the rice paddy and upland field were higher than that of the subsurface soil. These profiles are important for a comparison of the microbiological effects on hydraulic conductivities of soils at different depths.

A set of four samples was divided into two groups of 30 samples each. One group was treated with a water-soluble sterilizer (0.2% sodium azide) and another group was treated with a water-soluble nutrient (1% saccharose). The alternative wetting and drying procedure used by Frankenberger et al. (4) was adopted, as it was expected that both aerobic

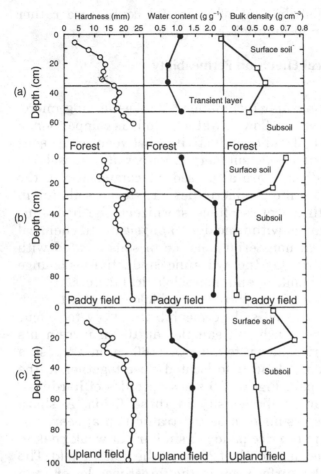

Figure 7.9 Profiles of hardness, water content, and bulk density of volcanic ash soil in (a) a forest, (b) a paddy field, and (c) an upland field. (After Miyazaki, T., Nakano, M., Shiozawa, S., and Imoto, H., *Trans. Jpn. Soc. Irrig. Drain. Reclam. Eng.* *155*:69–76 (1991). With permission.)

and anaerobic bacteria might become active during prolonged submergence.

Figure 7.10(a) shows the change in hydraulic conductivities of soils treated with water-soluble sterilizer, and Figure 7.10(b) shows the change in hydraulic conductivities of soils treated with water-soluble saccharose. The dimensionless

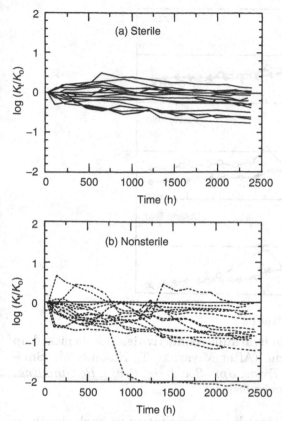

Figure 7.10 Logarithms of the ratio of hydraulic conductivities K_t to the initial hydraulic conductivities K_0 of samples treated with (a) water-soluble sterilizer and (b) water-soluble saccharose. (After Miyazaki, T., Nakano, M., Shiozawa, S., and Imoto, H., *Trans. Jpn. Soc. Irrig. Drain. Reclam. Eng. 155*:69–76 (1991). With permission.)

hydraulic conductivities in Figure 7.10 were obtained by dividing each hydraulic conductivity value by its initial value. The hydraulic conductivities of soils treated with nutrient water decreased more than those of soils treated with sterilized water, due to the microbiological effects of the nutrient water.

As shown in Figure 7.5–Figure 7.9, soil profiles may affect the change in hydraulic conductivities. Figure 7.11–Figure 7.14

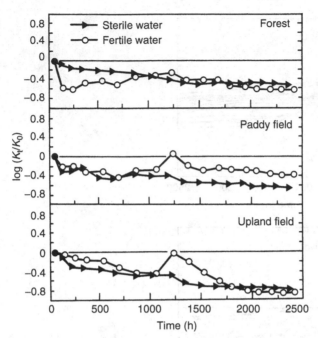

Figure 7.11 Changes in hydraulic conductivities of soils picked up from depths of 0 to 10 cm. (After Miyazaki, T., Nakano, M., Shiozawa, S., and Imoto, H., *Trans. Jpn. Soc. Irrig. Drain. Reclam. Eng.* *155*:69–76 (1991). With permission.)

show the change in hydraulic conductivities at each depth in forest soils, paddy field soils, and upland field soils. In the top soils of depth 0 to 10 cm (Figure 7.11), a gradual decrease occurred in all the samples, but there were no notable differences in the change between sterile and nutrient treatments. The decreased values in soils treated with sterile water are attributed to the changes of soil structure during treatment with sodium azide. The decreased values in soils treated with nutrient water are attributed to an increased bacteria count. In subsoils 20 to 30 cm deep (Figure 7.12), there were no notable differences in the change in hydraulic conductivities between sterile and nutrient treatments in forest soil and upland soil, whereas a big difference appeared in the paddy field soil. The decrease in hydraulic conductivity in the soil treated with nutrient water is attributed partly to an increase

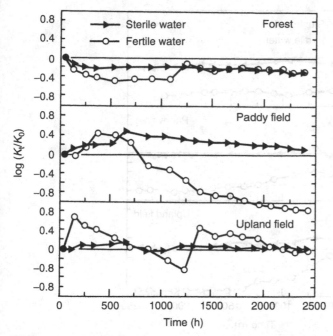

Figure 7.12 Changes in hydraulic conductivities of soils picked up from depths of 20 to 30 cm. (After Miyazaki, T., Nakano, M., Shiozawa, S., and Imoto, H., *Trans. Jpn. Soc. Irrig. Drain. Reclam. Eng.* 155:69–76 (1991). With permission.)

in microorganisms and partly to a reduction in iron accumulated at this particular depth in paddy fields. In subsoils 30 to 40 cm deep (Figure 7.13) and 50 to 60 cm deep (Figure 7.14), little change in hydraulic conductivities occurred in soils treated with sterile water, and an obvious decrease occurred in soils treated with nutrient water, probably due to the increase in microorganisms.

IV. EFFECTS OF MICROBIOLOGICAL FACTORS ON WATER FLOW IN CLAY LINERS OF LANDFILL SITES

Clay liners are used as contaminant barriers of solid waste disposal facilities. Kamon et al. (12) investigated the effects of the redox potential and the microbial activity on the hydraulic

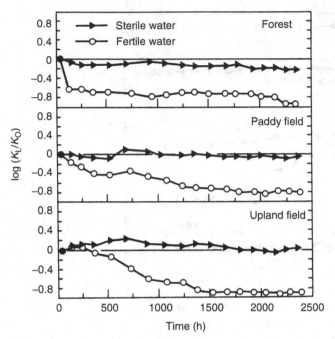

Figure 7.13 Changes in hydraulic conductivities of soils picked up from depths of 30 to 40 cm. (After Miyazaki, T., Nakano, M., Shiozawa, S., and Imoto, H., *Trans. Jpn. Soc. Irrig. Drain. Reclam. Eng.* 155:69–76 (1991). With permission.)

conductivity of Osaka marine clay used for landfill bottom liners. In their experiment, dithionite was used to maintain the soil system in a low Eh, i.e., reduced condition, and such nutrients as 1000 ppm of zinc acetate and 100 ppm of ammonium dihydrogenphosphate were used to supply bacteria in the soil with substrates. Osaka marine clay is composed of 45.8% clay, 52.1% silt, and 2.1% sand fractions, with 0.63% of low total organic carbon. They conducted the hydraulic conductivity test by mixing 20% quartz sand because this mixture is widely used to optimize the hydraulic conductivity of compacted clay liner in both laboratory researches and constructions of prototype liners. The mixed quartz sands were coated with Fe(III) to enhance the redox reaction during the hydraulic conductivity tests, in runs 2, 3, and 4 out of four runs.

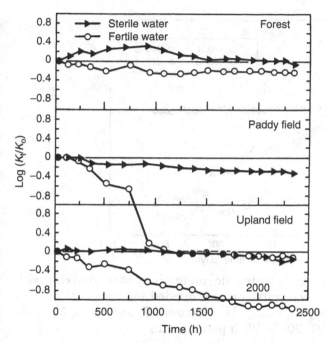

Figure 7.14   Changes in hydraulic conductivities of soils picked up from depths of 50 to 60 cm. (After Miyazaki, T., Nakano, M., Shiozawa, S., and Imoto, H., *Trans. Jpn. Soc. Irrig. Drain. Reclam. Eng.* 155:69–76 (1991). With permission.)

Kamon et al. (12), however, found that there were no obvious changes in the free swell index, the liquid limit, or the hydraulic conductivity of the marine clay when such strong reducing agents as dithionite and Fe-coating were used. On the other hand, they found that the hydraulic conductivities of the clay samples decreased as much as three orders, as shown in Figure 7.15, during 80 days when nutrients were applied for the growth of microorganisms. They concluded that both the formation of biofilm on the surface of the soil particles and anaerobic inorganic precipitation in the soil pores are responsible for the remarkable reduction in hydraulic conductivity. Finally, since clay liners are practically subject to reduced conditions, they concluded that these anaerobic microbial activities may provide rather preferable

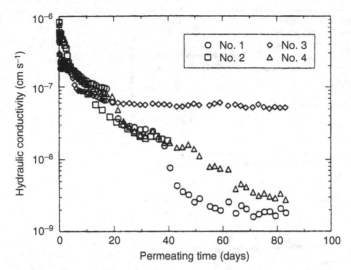

Figure 7.15   Change in hydraulic conductivity of marine clay with permeating time when nutrients are applied for microbial production. (After Kamon, M., Zhang, H., Katsumi, T., and Sawa, N., *Soils Found. 42*(6):79–91 (2002). With permission.)

effects on the hydraulic performance of marine clay as natural landfill liners at offshore landfill sites in Japan.

V. MICROBIAL CLOGGING MODELS

A. Microbial Clogging in Soils

There are two hypotheses to explain the mechanism of clogging in soils due to microbial growth resulting in a reduction in saturated hydraulic conductivities. The first is the *biofilm model*, where bacteria cover the pore walls of soil with biofilms in which the cells are intimately associated with a meshwork of exopolymers or glycocalyx. The second is the *microcolony model*, where the adherent bacteria exist dispersed on the solid surface, forming microcolonies that grow by consuming substrate and electron acceptor in the soil solution.

Harvey et al. (13) found that a majority of bacteria in contaminated and uncontaminated zones of an aquifer in Cape Cod, Massachusetts, were bound to the surfaces of soil particles less than 0.06 mm in diameter, and that microcolonies of between 10 and 100 bacteria were common features on these soil particle surfaces.

On the other hand, McCarty et al. (14) pointed out that "to adequately understand the kinetics of microbiological processes in groundwater, an understanding of biofilm kinetics and their application to groundwater system is necessary."

Vandevivere and Baveye (15) reported a saturated hydraulic conductivity reduction caused by aerobic bacteria in sand columns. They presented a relation between the value of K_s/K_{s0} versus biomass density in the soil column as shown in Figure 7.16, where K_s is the saturated hydraulic conductivity and K_{s0} is its initial value. The K_s/K_{s0} value decreased markedly with biomass density, almost exponentially. Through elaborate analysis of the experimental results, they confirmed the commonly observed fact that much of the reduction in hydraulic conductivity takes place at the inlet end of the sand column, associated with the formation of bacterial mat. Further, depending on the experimental evidence that bacteria did not form biofilms uniformly covering the pore

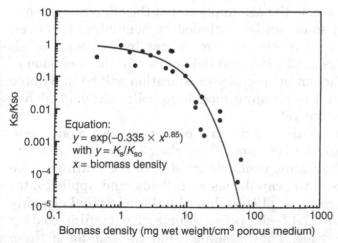

Figure 7.16 Relation between the values of K_s/K_{s0} versus biomass density in the soil column. (After Vandevivere, P. and Baveye, P., *Soil Sci. Soc. Am. J.* 56:1–13 (1992). With permission.)

walls, they concluded that the biofilm model would not be mechanistically appropriate to describe the accumulation of biomass in the pore space. They noted that this indication is applicable when the C/N ratio is normal (less than 39 in their experiment) but is not necessarily applicable when the C/N ratio is high (77 in their experiment). Finally, they postulated that the mechanism responsible for most of the reduction in hydraulic conductivity is accumulation of the cells themselves.

Several other studies have suggested that most microbes grow in colonies attached to solid surfaces rather than as attached uniform biofilms or free-floating organisms suspended in the liquid phase. Although the mechanism of microbial clogging in soils and resultant reduction of water permeability is still under investigation both experimentally and theoretically, it seems at present that the evidence supporting the microcolony concept is stronger than the evidence supporting the biofilm concept.

B. Microbial Clogging Models

Three typical models have been developed in recent years to describe the microbial clogging process in soils. All these models are based on the assumption that the effects of gaseous production in soils can be excluded or negligible. However, referring to the experimental evidences by Reynolds et al. (5), de Losanda et al. (16), and Seki et al. (6), the inclusion of gaseous production by microbial respiration will be inevitable in establishing the clogging models in soils, although it has not been achieved yet.

The first existing model is a biofilm-coating model, proposed originally by Ives and Pienvichitr (17) to describe the reduction of hydraulic conductivity due to the coating of the internal walls of the capillaries by colloids, and applied later by Vandevivere et al. (18) to describe the microbial clogging process. Figure 7.17 shows an illustrative biofilm coating in a capillary tube. They assumed that the bundle of these capillary tubes represents a porous material. By defining the

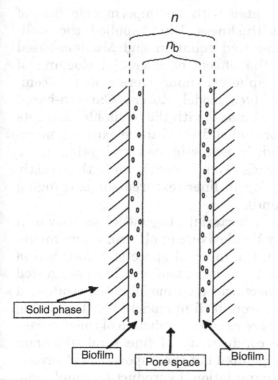

Figure 7.17 Schematic illustration of biofilm coating in a capillary tube.

biovolume ratio α as the bulk volume of biofilm per unit pore volume of the clean unclogged porous medium with,

$$\alpha = \frac{n - n_b}{n} \tag{7.1}$$

where n is the initial porosity and n_b the porosity after the coating of capillary tubes by biofilms, Vandevivere et al. (18) recasted the resultant saturated hydraulic conductivity of this materials as

$$K_s = K_{s0}(1 - \alpha)^{3-2p} \tag{7.2}$$

where p is a parameter larger than or equal to zero.

The second model was proposed by Taylor et al. (19), who assumed that a biofilm is developed in such a way that all the uniform spheres constituting the porous medium are regularly packed and coated with an impermeable film of biomass of a constant thickness. They applied the well-known Kozeny–Carman-based equation and Mualem-based equation to formulate the effects of microbial clogging of these saturated uniform spheres. Vandevivere et al. (18) compared the agreements of Ives' model, Kozeny–Carman-based model, and Mualem-based model with the available data sets in the literature, and concluded that all these existing microbial clogging models slightly overestimate the clogging in the coarse-textured glass beads (1-mm beads), while they vastly underestimate the clogging in finer-textured sands (ranging from 0.7- to 0.09-mm sand).

The third model is a bacterial plug model as shown in Figure 7.18, proposed by Vandevivere et al. (18), where microorganisms form plugs in the internal space of porous media instead of (or in addition to) forming biofilms. They suggested that, even though this bacterial plug model is too crude and simplistic, the alteration from biofilm model to bacterial plug model brought about a more reliable prediction of the decrease in saturated hydraulic conductivity of fine sand (0.09-mm sand). In addition, they pointed out the importance of involving the effects of cell accumulation, byproduct accumulation,

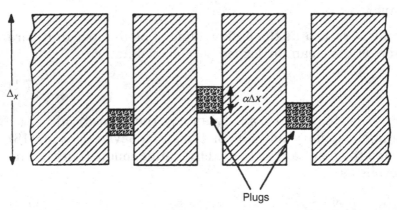

Figure 7.18 Schematic illustration of a simple clogging model based on the formation of bacterial plugs in the lumen of uniform, cylindrical capillaries. (After Vandevivere, P., Baveye, P., Sanchez de Lozada, D., and DeLeo, P., *Water Resour. Res. 31*:2173–2180 (1995). With permission.)

gas production, and predation in the microbial clogging processes.

C. Colony-Enveloping Model

As Vandevivere et al. (18) pointed out, mathematical models assuming the microorganisms to be present only in biofilms uniformly coating the solid surfaces have oversimplified the pore-scale geometrical configurations of clogged soils. The plug model given in Figure 7.18 was a possible alternative but still gave a limited success.

Seki and Miyazaki (20) made a mathematical model for biological clogging by taking into account the effect of microbial distribution on each soil particle. They introduced a colony-enveloping space concept as shown in Figure 7.19 to describe the microbial colony formation and to relate it to water flow in soils. The space was assumed to form a film with uniform thickness of L_b that was identical to the maximum colony thickness among existing colonies. The biofilm formation around the soil particles is involved in this model as an extreme case when there are so many microbial colonies

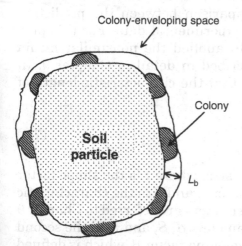

Figure 7.19 Schematic illustration of colony-enveloping space model whose thickness, L_b, is equal to the maximum colony thickness around a soil particle. (After Seki, K. and Miyazaki, T., *Water Resour. Res.* 37(12):2995–2999 (2001). With permission.)

on each soil particle that they are completely connected to each other. Saturated water flow in the soil where the colony-enveloping spaces are thus formed may be definitely affected by the pore throat size between each soil particle and by the pressure gradient.

A convenient and reliable relation between the saturated water flow and the pore throat size in porous media was given by Ewing and Gupta (21), who approximated that the flux of water through a circular pore throat is proportional to the third power of the throat diameter. Based on this approximation, Seki and Miyazaki (20) related the saturated hydraulic conductivity of the clogged soil and the pore throat size as

$$\frac{d - 2L_b}{d} = \left(\frac{K_s}{K_{s0}}\right)^{1/3} \tag{7.3}$$

where d is the initial pore throat size, K_s is the saturated hydraulic conductivity of the clogged porous medium, and K_{s0} is the saturated hydraulic conductivity of the clean porous medium. It is more preferable to relate the saturated hydraulic conductivity ratio K_s/K_{s0} to the biovolume ratio α defined by (7.1) for the comparison between the prediction by the model and available experimental data. For this purpose, Seki and Miyazaki (20) applied the nonsimilar media concept (NSMC) model, described in detail in Chapter 9. In the first step, they assumed that the characteristic length of the pore phase d, given by

$$d = \left\{\left(\frac{\tau}{1 - n}\right)^{1/3} - 1\right\}S \tag{7.4}$$

is identical to the pore throat size, where τ is the shape factor of the solid phase, n is the porosity, and S is the characteristic length of the solid phase. The readers are referred to Chapter 9 for the definitions of the parameters d, S, and τ. In the second step, they introduced the enveloping factor β, which is defined as the bulk volume of biomass per unit volume of colony-enveloping space, in the form of

$$\beta = \frac{\alpha n}{1-n}\left\{\left(1+\frac{2L_b}{S}\right)^3 - 1\right\}^{-1}, \quad 0 < \beta \le 1 \tag{7.5}$$

Equation (7.5) was easily obtained geometrically by comparing the volume ratio of biomass to solid clean particles in terms of biovolume ratio α as

$$\frac{V_b}{V_s} = \frac{\alpha n}{1-n} \tag{7.6}$$

and in terms of enveloping factor β as

$$\frac{V_b}{V_s} = \frac{\beta V_f}{V_s} \tag{7.7}$$

where V_b is the biomass volume, V_s is the volume of the clean solid phase, and V_f is the volume of the colony-enveloping space. When many colonies are forming biofilms around the soil particles, β can be set as equal to unity.

In their third step, Seki and Miyazaki (20) obtained empirical relationships between L_b and β for each range of beads sizes by using the data of Cunningham et al. (22) and Vandevivere and Baveye (23), as shown in Table 7.3. Once the empirical relationship between L_b and β is found, and when the values n and S are given, we can determine the unknown L_b value as a function of the biovolume ratio α by solving the simultaneous equations (Equation (7.5)) and one of the equations in Table 7.3. It is plausible to assume that the

Table 7.3 Empirical Equations of β as Functions of L_b for Each Size Fraction of Beads Calculated from the Data in Ref. (18)

Sample	Fitted equation
1-mm beads	$\beta = 11.383 L_b^{-0.5566}$
0.7-mm beads	$\beta = 0.0004 L_b^2 - 0.0377 L_b + 1.3160$
0.54-mm beads	$\beta = 0.0005 L_b^2 - 0.0343 L_b + 0.9190$
0.12-mm beads	$\beta = 0.0009 L_b^2 - 0.0462 L_b + 0.8748$
0.09-mm sand	$\beta = -0.0044 L_b + 0.1945$

Source: Seki, K. and Miyazaki, T., *Water Resour. Res.* 37(12):2995–2999 (2001). With permission.

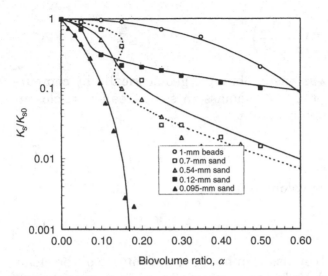

Biovolume ratio, α

Figure 7.20 Saturated hydraulic conductivity ratio as a function of biovolume ratio. Measured values (symbols) are from Ref. (18) and predicted values (lines) are based on the colony-enveloping space model. (After Seki, K. and Miyazaki, T., *Water Resour. Res.* *37*(12):2995–2999 (2001). With permission.)

characteristic length of solid phase, S, is equal to the representative particle size of the beads.

In the final step, they calculated the saturated hydraulic conductivity ratio K_s/K_{s0} as a function of the biovolume ratio α through Equation (7.3) and obtained Figure 7.20, where good agreements between published experimental data and theoretically calculated values were obtained. It is notable that by introducing the enveloping factor β in the colony-enveloping space concept, the abrupt decreases in the K_s/K_{s0} values against α were successfully reproduced.

VI. MICROBIAL GROWTH AND SUBSTRATE TRANSPORT MODEL

A. Microcolony Concept

The internal mechanisms of direct and indirect microbial effects on hydraulic properties of soils are still vague, as mentioned above. Although there are many concepts and models describing the relation between microbial growth and hydraulic properties of soils as reviewed by Baveye and Valocchi (24), all of them are still under investigation.

Based on observations by Harvey et al. (13), who confirmed that more than 95% of the total biomass was attached to particle surfaces, Molz et al. (25) adopted the microcolony model in their simulation rather than the biofilm model. Although they assumed as a condition of their calculations that the increased number of microcolonies in a porous medium does not affect the saturated hydraulic conductivity value, it is likely that microbial growth in the form of microcolonies reduces the permeability of natural soils. The microcolony concept of aerobic microorganisms proposed by them may help us understand how microbial growth is related to the transport of substrate and oxygen in soils. The following description is based largely on their work.

Figure 7.21 shows the microcolony model proposed by Molz et al. (25). Here, a typical microcolony is modeled using a uniform cylindrical plate of radius r_c and thickness τ. A diffusion boundary layer of thickness δ separates the pore bulk liquid from the microcolony. Each microcolony is assumed to occupy separate sites on solid surfaces. Substrate and oxygen are assumed to be supplied by molecular diffusion through the diffusion boundary layer from bulk solution to

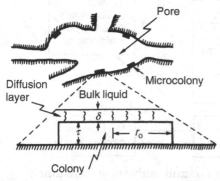

Figure 7.21 Microcolony model. (After Molz, F. J., Widdowson, M. A., and Benefield, L. D., *Water Resour. Res.* 22(8):1207–1216 (1986). With permission.)

each colony. The diffusion fluxes of substrate and oxygen must be identical with the amounts consumed in each microcolony.

A marked feature of their microcolony model is that the biomass growth in a porous medium is not accounted for by the increases in each microcolony but by the increased number of microcolonies, even though the diffusion layer model above each microcolony (Figure 7.21) is adopted. This manipulation is slightly confusing but makes for simpler mathematical treatment.

B. Conservation and Production of Biomass in a Microcolony

1. Flow Rate Chart of Material

Figure 7.22 shows a flowchart of the oxygen and substrate consumed by a microcolony during conservation and production of biomass. Note that the organic removal rate is identical to the rate of substrate utilization per colony. In Figure 7.22, the rate of substrate consumption (r_s) is composed of the rate of gross biomass production (Yr_s) and the rate of substrate

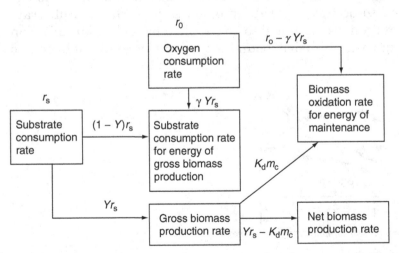

Figure 7.22 Flowchart of oxygen and substrate consumed by a microcolony. (Modified from Molz, F. J., Widdowson, M. A., and Benefield, L. D., *Water Resour. Res.* 22(8):1207–1216 (1986). With permission.)

consumption for the supply of energy of gross biomass production $[(1 - Y)r_s]$. Oxygen is consumed as an electron acceptor oxidizing both substrate and biomass. The rate of oxygen consumption (r_0) is composed of the rate of oxidization of substrate $(\gamma Y r_s)$, which provides the energy for gross biomass production, and the rate of oxidation of biomass $(r_0 - \gamma Y r_s)$, which provides the energy for maintenance of biomass. A part of the gross biomass produced is oxidized to provide the energy for maintenance of biomass (whose rate is $K_d m_c$), and the remaining part is the net biomass (whose rate is $Y r_s - K_d m_c$). All the notations in Figure 7.22 are defined in the following sections.

2. Monod Equation

When the only variable in a given system is substrate concentration S (mass volume^{-1}), biomass production rate R (mass time^{-1}) is related to S by the Monod equation,

$$R = R_{ms}\frac{S}{K_s + S} \tag{7.8}$$

where R_{ms} is the potential biomass production rate provided that sufficient substrate is supplied (mass time^{-1}), and K_s is the substrate saturation constant (mass volume^{-1}). Similarly, when the only variable in the system is oxygen concentration O (mass volume^{-1}), R is related to O by

$$R = R_{mo}\frac{O}{K_0 + O} \tag{7.9}$$

where R_{mo} is the potential biomass production rate provided that sufficient oxygen is supplied (mass time^{-1}) and K_o is the oxygen saturation constant (mass volume^{-1}). Figure 7.23 gives an outline of the Monod equation when the only variable is S. The modified Monod equation where both substrate concentration and oxygen concentration change is given by (25)

$$R = R_m\frac{S}{K_s + S}\frac{O}{K_0 + O} \tag{7.10}$$

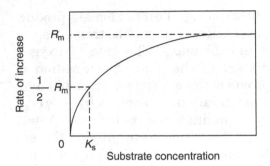

Figure 7.23 Conceptual curve of the Monod equation.

where R_m is the potential biomass production rate (mass time^{-1}) when sufficient substrate and oxygen are supplied.

Application of Equation (7.10) to the gross biomass production rate per microcolony yields a modified Monod relationship for microbial growth kinetics (25):

$$Yr_s = \mu_m m_c \frac{s}{K_s + s} \frac{o}{K_o + o} \qquad (7.11)$$

where Y is the yield coefficient for microorganisms (dimensionless), r_s is the rate of substrate utilization per colony (per mass colony per time), the product Yr_s is the gross biomass production rate, μ_m is the maximum specific growth rate of microorganisms (mass mass^{-1} time^{-1}), m_c is the mass of microorganisms per colony (mass colony^{-1}), s is the substrate concentration within a colony (mass volume^{-1}), and o is the oxygen concentration within a colony (mass volume^{-1}). The product of μ_m and m_c is the potential biomass production rate when sufficient substrate and oxygen are supplied.

Mass balances of substrate and biomass give other quantitative relations in Figure 7.22. For example, the remainder

$$r_s - Yr_s \qquad (7.12)$$

gives the substrate oxidation rate that is required to provide the energy of biomass production. The main part of the gross biomass produced contributes to produce net biomass, while the remainder contributes to maintaining the biomass itself. Assuming that the biomass oxidation rate for maintenance energy is proportional to the biomass per colony m_c, the biomass oxidation rate is given by

$$K_d m_c \qquad (7.13)$$

where K_d is the microbial decay coefficient (mass mass^{-1} time^{-1}) (25). The remainder of the gross biomass,

$$Yr_s - K_d m_c \qquad (7.14)$$

gives the net biomass production rate.

Oxygen is consumed to provide the energy for biomass production. Denoting the oxygen use coefficient for synthesis of biomass by γ (dimensionless), the oxygen consumption rate as an electron acceptor is given by

$$\gamma Yr_s \qquad (7.15)$$

The remainder,

$$r_0 - \gamma Yr_s \qquad (7.16)$$

gives the oxygen consumption rate required to provide the energy for maintenance of biomass by its oxidation. The Monod equation also gives the same oxygen consumption rate independently as

$$r_0 - \gamma Yr_s = \alpha K_d m_c \frac{o}{K'_0 + o} \qquad (7.17)$$

where α is called the oxygen use coefficient for energy of maintenance (25) and K'_0 is the oxygen saturation constant for decay.

3. Fluxes in Diffusion Layer

Independent of the Monod relation, the substrate consumption rate r_s and the oxygen consumption rate r_0 per microcolony are both estimated by assuming that they are equal to the diffusion fluxes in the diffusion layer between the bulk liquid and the microcolony in the model (Figure 7.20). The diffusion equations are given by

$$r_s = D_{sb} \frac{S - s}{\delta} \pi r_c^2 \qquad\qquad (7.18)$$

$$r_o = D_{ob} \frac{O - o}{\delta} \pi r_c^2 \qquad\qquad (7.19)$$

where D_{sb} and D_{ob} are diffusion coefficients in the boundary layer for substrate and oxygen, respectively; S is the substrate concentration in the bulk liquid (mass volume^{-1}); O is the oxygen concentration in the bulk liquid (mass volume^{-1}); and δ is the thickness of the diffusion boundary layer.

4. Advection–Dispersion Equations

One-dimensional liquid flow in saturated soils is denoted by Darcy's equation. Figure 7.24 is the one-dimensional control volume, in which a large number of soil particles and microcolonies are contained. Microcolonies decrease both substrate and oxygen values in the advection–dispersion equation.

On the same scale, using Darcy's equation, substrate flow and oxygen flow in saturated soils are described by advection–dispersion equations. The advection flux of substrate, q_{as} (mass length^{-2} time^{-1}), and the advection flux of oxygen, q_{ao} (mass length^{-2} time^{-1}), are given by

$$q_{as} = n v_x S \qquad\qquad (7.20)$$

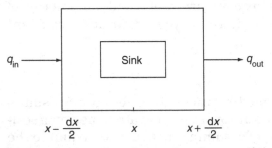

Figure 7.24 One-dimensional control volume in which a large number of soil particles and microcolonies are contained as a sink of substrate and oxygen.

$$q_{ao} = nv_x O \tag{7.21}$$

respectively, where n is the porosity (dimensionless) and v_x is the average velocity of the pore fluid (length time^{-1}). The dispersion flux of substrate, q_{ds} (mass length^{-2} time^{-1}), and that of oxygen, q_{do} (mass length^{-2} time^{-1}), are given by

$$q_{ds} = -nD_s \frac{\partial S}{\partial x} \tag{7.22}$$

$$q_{do} = -nD_o \frac{\partial O}{\partial x} \tag{7.23}$$

where D_s and D_o are the dispersion coefficients of substrate and oxygen (length^2 time^{-1}), respectively. The total flux of substrate q_s is then given by

$$q_s = q_{as} + q_{ds} \tag{7.24}$$

and the total flux of oxygen q_o is given by

$$q_o = q_{ao} + q_{do} \tag{7.25}$$

The sink term in the advection–dispersion equation is composed of the consumptions of substrate and oxygen by microbes in the control volume (Figure 7.24). Denoting the number of microcolonies contained in a unit volume by N_c, the consumption rate of substrate and oxygen in a unit volume is given by $N_c r_s$ (mass volume^{-1} time^{-1}) and $N_c r_o$ (mass volume^{-1} time^{-1}), respectively.

The amount of substrate per unit volume S_{ts} (mass volume^{-1}) is composed of two fractions, one contained in bulk liquid and another absorbed by solid surfaces, such that

$$S_{ts} = nS + S_a A_s \tag{7.26}$$

where S_a is the mass of substrate absorbed per unit solid surface (mass length^{-2}) and A_s is the effective specific surface of solid surfaces (length^2 length^{-3}). In the same manner, the amount of oxygen per unit volume O_{to} (mass volume^{-1}) is given by

$$O_{to} = nO + O_a A_s \tag{7.27}$$

where O_a is the mass of oxygen absorbed per unit solid surface
(mass length^{-2}).

The advection–dispersion equations of substrate and oxygen are thus given by

$$\frac{\partial S_{ts}}{\partial t} = -\frac{\partial q_s}{\partial X} - N_c r_s \tag{7.28}$$

and

$$\frac{\partial O_{ts}}{\partial t} = -\frac{\partial q_o}{\partial X} - N_c r_o \tag{7.29}$$

respectively.

5.   Solution of Advection–Dispersion Equations

Numerical solutions of advection–dispersion equations (7.28)
and (7.29) were obtained by Molz et al. (25) using an iterative
numerical method. In their simulation, the initial substrate
concentration S and the initial oxygen concentrations O in
the bulk liquid were 0.5 mg l^{-1}, and the mass of substrate
and oxygen absorbed by solid surfaces was assumed to be
proportional to their mass in the bulk liquid.

Figure 7.25 is an example of their simulation showing the
profiles of substrate concentration S, oxygen concentration O,
and colony density N_c after 4 days of liquid percolation. The
initial condition in their calculation was given such that the
substrate concentration at the boundary $X = 0$ was increased
instantaneously from 0 to 15 mg l^{-1} at time zero, and the
retardation factors f_{rs} and f_{ro} were subjected to the calculations. Other values of parameters used by Molz et al. are listed
in Table 7.4. These were selected appropriately from the literature and their own estimations for purposes of illustration.
The dimensions of the parameters for mass, length, volume,
and time in the table were chosen arbitrarily based on the
magnitudes of the values.

In Figure 7.25, colony density N_c was calculated assuming that the average colony size and mass remain constant

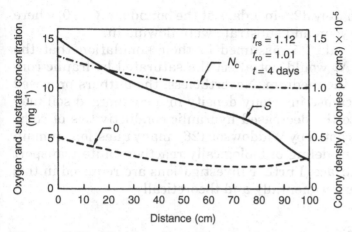

Figure 7.25 Simulated profiles of substrate, oxygen, and colony density after percolation of liquid for 4 days. (After Molz, F. J., Widdowson, M. A., and Benefield, L. D., *Water Resour. Res.* 22(8):1207–1216 (1986). With permission.)

Table 7.4 Parameters Used in the Model Calculation by Molz et al. (25)

Parameter	Value	Remark
μ_m (day^{-1})	4.34	Maximum specific growth rate of microorganisms
Y	0.278	Yield coefficient for microorganisms
K_d (day^{-1})	0.02	Microbial decay coefficient
m_c (ng)	0.03534	Mass of microorganisms per microcolony
D_{sb} (cm^2 day^{-1})	0.06	Diffusion coefficient of substrate
D_{ob} (cm^2 day^{-1})	0.71	Diffusion coefficient of oxygen
K_o' (mg cm^{-3})	0.00077	Oxygen saturation constant for decay
K_s (mg cm^{-3})	0.120	Substrate saturation constant
K_o (mg cm^{-3})	0.00077	Oxygen saturation constant
α	0.0402	Oxygen use coefficient for energy of maintenance
γ	1.4	Oxygen use coefficient for synthesis of biomass
δ (cm)	0.05	Thickness of diffusion boundary layer

even though the individual colony mass varies, which enable us to convert the increased biomass in a unit volume into an increased number of colonies per unit volume N_c. It is noted in Figure 7.25 that the number of colonies increased by approximately 32% in 4 days at the boundary $X = 0$, where sufficient oxygen and substrate were flowing in.

Molz et al. (25) assumed in their simulation that the increase in N_c would not affect the saturated hydraulic conductivity of the medium. Nevertheless, the authors proposed that this increase in colony density in a submerged soil may be related to the decreased hydraulic conductivities of soils. However, as noted by Widdowson (26), many questions remain concerning modeling of biologically reactive solute transport in the subsurface. Further investigations are required in this field, both experimentally and theoretically.

REFERENCES

1. Allison, L. E., Effect of microorganisms on permeability of soil under prolonged submergence, *Soil Sci.* *63*:439–450 (1947).
2. Martin, J. P., Microorganisms and soil aggregation. I. Origin and nature of some of the aggregating substances, *Soil Sci.* *59*:163–174 (1945).
3. Poulovassilis, A., The changeability of the hydraulic conductivity of saturated soil samples, *Soil Sci.* *113*(29):81–87 (1972).
4. Frankenberger, W. T., Jr., Troeh, F., and Dumenil, L. C., Bacterial effects on hydraulic conductivity of soils, *Soil Sci. Soc. Am. J.* *43*:333–338 (1979).
5. Reynolds, W. D., Brown, D. A., Mathur, S. P., and Overend, R. P., Effect of in-situ gas accumulation on the hydraulic conductivity of peat, *Soil Sci.* *153*(5): 397–408 (1992).
6. Seki, K., Miyazaki, T., and Nakano, M., Effects of microorganisms on hydraulic conductivity decrease in infiltration, *Eur. J. Soil Sci.* *49*:231–236 (1998).
7. Motomura, S., Dynamic behavior of ferrous iron in paddy soils, *Bull. Natl. Inst. Agric. Sci. Ser. B 21*:1–114 (1969).
8. Mitsuchi, M., Features of paddy soils originated in Andosols, *Jpn. J. Soil Sci. Plant Nutr. 41*(8):307–313 (1970).
9. Ishizawa, S. and Toyoda, H., Microflora of Japanese soils, *Bull. Natl. Inst. Agric. Sci. Ser. B 14*:203–284 (1964).
10. Fujikawa, T., Miyazaki, T., and Imoto, H., Study on behavior of CO_2 and O_2 gas in soil with a hard pan, *Trans. Jpn. Soc. Irrig. Drain. Reclam. Eng. 71*(3):111–118 (2003).

11. Miyazaki, T., Nakano, M., Shiozawa, S., and Imoto, H., The effects of microorganisms on the permeability of soils, *Trans. Jpn. Soc. Irrig. Drain. Reclam. Eng. 155*:69–76 (1991).

12. Kamon, M., Zhang, H., Katsumi, T., and Sawa, N., Redox effect on the hydraulic conductivity of clay liner, *Soils Found. 42*(6): 79–91 (2002).

13. Harvey, R. W., Smith, R. L., and Jeorge, L., Effect of organic contamination upon microbial distributions and heterotrophic uptake in a Cape Cod, Mass., aquifer, *Appl. Environ. Microbiol. 48*:1197–1202 (1984).

14. McCarty, P. L., Rittmann, B. E., and Bouwer, E. J., Microbiological processes affecting chemical transformations in groundwater, in *Groundwater Pollution Microbiology*, G. Bitton and C. P. Gerba, Eds., Wiley, New York, pp. 89–115 (1984).

15. Vandevivere, P. and Baveye, P., Saturated hydraulic conductivity reduction caused by aerobic bacteria in sand columns, *Soil Sci. Soc. Am. J. 56*:1–13 (1992).

16. de Losanda, D. S., Vandevivere, P., and Zinder, S., Decrease of the hydraulic conductivity of sand columns by Methanosarcina barkeri, *World J. Microbiol. Biotechnol. 10*:325–333 (1994).

17. Ives, K. J. and Pienvichitr, V., Kinetics of the filtration of dilute suspensions, *Chem. Eng. Sci. 20*:965–973 (1965).

18. Vandevivere, P., Baveye, P., Sanchez de Lozada, D., and DeLeo, P., Microbial clogging of saturated soils and aquifer materials: evaluation of mathematical models, *Water Resour. Res. 31*: 2173–2180 (1995).

19. Taylor, S. W., Milly, P. C. D., and Jaffe, P. R., Biofilm growth and the related changes in the physical properties of a porous medium, 2, Permeability, *Water Resour. Res. 26*:2161–2169 (1990).

20. Seki, K. and Miyazaki, T., A mathematical model for biological clogging of uniform porous media, *Water Resour. Res. 37*(12): 2995–2999 (2001).

21. Ewing, R. and Gupta, S. C., Pore-scale network modeling of compaction and filtration during surface sealing, *Soil Sci. Soc. Am. J. 58*:712–720 (1994).

22. Cunningham, A. B., Characlis, W. G., Abedeen, F., and Crawford, D., Influence of biofilm accumulation on porous media hydrodynamics, *Environ. Sci. Technol. 25*:1305–1311 (1991).

23. Vandevivere, P. and Baveye, P., Relationship between transport of bacteria and their clogging efficiency in sand columns, *Appl. Environ. Mi;* Reynolds, W. D., Brown, D. A., Mathur, S. P.,

and Overend, R. P., Effect of in-situ gas accumulation on the hydraulic conductivity of peat, *Soil Sci. 153*(5):397–408 (1992).

24. Baveye, P. and Valocchi, A., An evaluation of mathematical models of the transport of biologically reacting solutes in saturated soils and aquifers, *Water Resour. Res. 25*(6):1413–1421 (1989).

25. Molz, F. J., Widdowson, M. A., and Benefield, L. D., Simulation of microbial growth dynamics coupled to nutrient and oxygen transport in porous media, *Water Resour. Res. 22*(8):1207–1216 (1986).

26. Widdowson, M. A., Comment on "An evaluation of mathematical models of the transport of biologically reacting solutes in saturated soils and aquifers" by Philippe Baveye and Albert Valocchi, *Water Resour. Res. 27*(6):1375–1378 (1991).

8

Water Regimes in Fields with Vegetation

SHUICHI HASEGAWA
Division of Environmental Resources,
Hokkaido University,
Sapporo, Hokkaido, Japan

TATSUAKI KASUBUCHI
Department of Bioenvironment, Yamagata University,
Tsuruoka, Yamagata, Japan

TSUYOSHI MIYAZAKI
Department of Biological and Environmental Engineering,
The University of Tokyo, Japan

I. CHARACTERISTICS OF WATER FLOW
IN PADDY RICE FIELDS

A. Water Flow in Unsaturated Subsoil
in a Paddy Field

It is sort of amazing that, in paddy rice fields covered with surface irrigation water, the subsoils are continuously kept in unsaturated conditions, even when the downward water flows last for months.

Hamada and Miyazaki (1) measured the vertical soil moisture distributions in an alluvial fan in Japan, where many paddy fields were extending, by using a neutron moisture meter. The result in the irrigation period measured on July 27, 1998, and that in the nonirrigation period measured on January 25, 1999, are shown in Figure 8.1. The average volumetric water content in the subsoil from a depth of 1 m down to a depth of 13 m was 22% (±4%) both in nonirrigation periods and in even irrigation periods. The average volumetric water contents under the groundwater tables, where the soils were saturated, were 47% (±4%) and, therefore, the degree of saturation in the unsaturated subsoil was estimated to be about 47%.

Hamada and Miyazaki (1) succeeded in confirming the existence of the downward unsaturated water flow during the irrigation period by applying the quasipartition law of

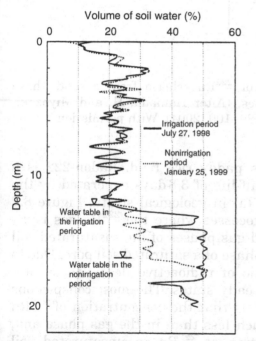

Figure 8.1 Vertical soil water profiles in irrigated and nonirrigated periods in a paddy rice field in Japan. (After Hamada, H. and Miyazaki, T., *J. Environ. Radioact.* 71:89–100 (2004). With permission.)

Unsaturated soil

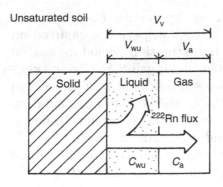

Saturated soil

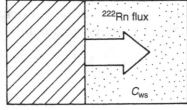

V : Volume
C : ^{222}Rn concentration

Figure 8.2 Steady state of ^{222}Rn release from a solid phase to liquid and gaseous phases. (After Hamada, H. and Miyazaki, T., *J. Environ. Radioact.* 71:89–100 (2004). With permission.)

radon-222 (^{222}Rn) in this paddy rice field. Radon-222 is a radioactive gas with a half-life of 3.8 days generated by the decay of radium-226 (^{226}Ra) in geological strata. Figure 8.2 gives conceptually the processes where the ^{222}Rn gas is released into the liquid and gas phases of an unsaturated soil pore, and into the liquid phase of a saturated soil pore. Due to the peculiar characteristic of radioactive decay, these processes are regarded as steady states. The most conspicuous behavior of ^{222}Rn in soils is that the concentration of ^{222}Rn in the liquid phase is much less than in the gas phase and, therefore, the concentration of ^{222}Rn in unsaturated soil water, C_{wu}, is less than that in saturated soil water, C_{ws}. In addition, the concentration ratio, R, defined by the ratio of C_{wu} to C_{ws}

$$R = \frac{C_{wu}}{C_{ws}} = \frac{100 D_w}{100 - (1 - D_w) S_r} \tag{8.1}$$

provides the estimation of the degree of saturation, S_r, at the ambient soil zone, where D_w is the partitioning coefficient approximated as a function of temperature T (°C) by Noguchi (2) as

$$D_w = \frac{9.12}{17.0 + T} \tag{8.2}$$

Whenever we sample fresh groundwater, ^{222}Rn is contained in it to a greater or less extent. If the concentration of ^{222}Rn of the fresh groundwater is relatively high, the water has not been mixed with adjacent unsaturated pore water, while if the concentration is relatively low, the water has been mixed with adjacent unsaturated pore water. Figure 8.3 shows the change in ^{222}Rn concentration of the groundwater under a paddy rice field during 2 years in the irrigation period (A, B, F) and nonirrigation periods (C, D, E, G). The groundwater level is also shown where it is higher during the irrigation periods and lower during the nonirrigation periods. It is very clearly shown that the ^{222}Rn concentration of the groundwater was less during the irrigation periods due to mixing with unsaturated soil water flowing down into the groundwater. During the nonirrigation period, downward unsaturated water flow hardly exists in deep soil zone and, therefore, the ^{222}Rn concentration of the groundwater was high.

Since the degree of saturation in the unsaturated subsurface soil zone calculated by Equation (8.1) was 49%, it is safe to say that the agreement of the degree of saturation determined by the ^{222}Rn concentration and by the neutron moisture meter was excellent.

B. Feature of the Matric Potential Profile in a Paddy Field

Takagi (3) analyzed the vertical downward steady flow of water through a two-layered soil of less permeable top layer and more

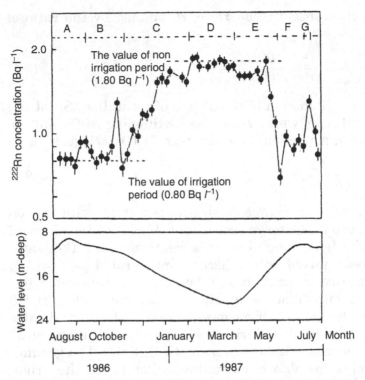

Figure 8.3 Changes in ^{222}Rn concentration in the groundwater, corresponding to its fluctuation. Bars are the counting errors. (After Hamada, H. and Miyazaki, T., *J. Environ. Radioact.* 71:89–100 (2004). With permission.)

permeable sublayer and emphasized that there exists a peculiar zone in the sublayer where the matric potential gradient is zero and, consequently, the downward water flux is equal to the unsaturated hydraulic conductivity. Yamazaki (4) had found this peculiarity of unsaturated downward water flow earlier. Srinilta et al. (5) also analyzed the steady downward flow of water through a two-layer soil based on Darcy's equation, and concluded that a zone of constant matric potential never exists. Figure 8.4 schematically gives the point at issue whether the matric potential gradient can be zero or larger than zero in the upper zone of the sublayer from L_1 to L_2. Note that, in Figure 8.4, the water pressure at the land surface is kept at a constant positive value due to continuous irriga-

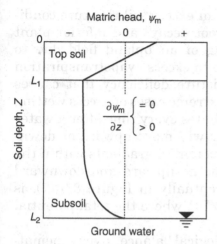

Figure 8.4 A question schematically shown whether the matric potential of the subsoil in a paddy rice field under irrigation is constant or not.

tion. Iwata et al. (6) summarized the issue that the gradient of matric potential $d\phi_m/dz$, having positive values in this coordinate system, always decreases for the upward direction and gradually approaches zero. Accordingly, when the unsaturated subsoil zone is long enough, the zone with constant pressure $(d\phi_m/dz = 0)$ will appear in the upper part of the sublayer.

Figure 8.1 agrees well with the conclusion given above in an actual field because the constant water content profile in the unsaturated zone from a depth of 1 m down to a depth of 13 m will provide the constant matric potential in this zone. In many paddy rice fields in the world, similar situations will be seen except when the groundwater level is very close to the land surface.

II. CHARACTERISTICS OF WATER FLOW IN UPLAND AGRICULTURAL FIELDS

A. Water Flow in Unsaturated Subsoil in an Upland Field

The most favorable soil condition in upland agricultural fields is available when the moisture condition is moderate. Poor drainage or too much water supply by natural rains or by artificial irrigations may cause an excess soil moisture condition in the field, resulting in root decays and inferior plant growth. Contrarily, over drying of an upland field due to shortage of water supply or due to excess evapotranspiration (ET) will bring about soil moisture deficiency that causes plant wilting. Except for these extreme cases of excess wetting or excess drying, water redistributes every time after a water supply, during which soil water will move upward or downward in proportion to the total potential gradients within the soil. The border plane between the upward and downward water movements, shown conceptually in Figure 8.5 (7), is defined as the zero flux plane, ZFP, where the total potential gradient dH/dz is zero.

Figure 8.6 shows the hydrological balance, given schematically by Sharma et al. (8), in a field with vegetation located at a semiarid region. The root zone reached at a depth of about 10 m and the groundwater table was at a depth of about 18 m. By

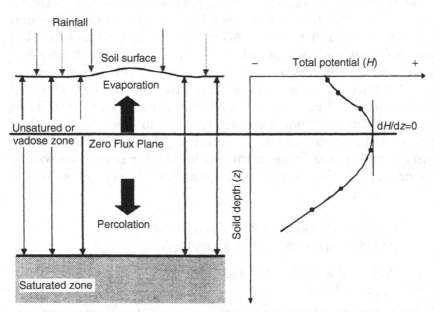

Figure 8.5 Zero flux plane concept. (After Khalil, M., Sakai, M., Mizoguchi, M., and Miyazaki, T., *J. Jpn. Soc. Soil Phys. 95*:75–90 (2003). With permission.)

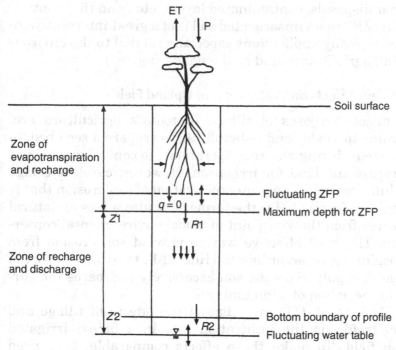

Figure 8.6 Schematic representation of the hydrological balance used in the analysis. Here, $Z_1 = 10$ m and $Z_2 = 18$ m are arbitrary depths. Z_1 is the maximum depth for the zero flux plane (ZFP) and recharge below this depth is termed R_1. Z_2 is an arbitrary depth just above the water table and recharge below this depth is termed R_2. (After Sharma, M. L., Bari, M., and Byrne, J., *Hydrol. Process.* 5:383–389 (1991). With permission.)

monitoring the moisture content for 3 years with neutron probes, and by converting the moisture content profiles into the total potential profiles, they found that the ZFP moves from very shallow zone down to a depth of 10 m according to the water balance among precipitation (P), ET, and soil water redistribution. It is noted that water flow in upland agricultural fields may also be characterized to a greater or less extent similar to the case as shown in Figure 8.5 where the ZFP exists.

Khalil et al. (7) pointed out that it is very important to maintain the ZFP higher than the salt accumulated subsoil, nuclear disposals, contaminated layers, etc., and that controlling the ZFP in an unsaturated soil is of a great interest where there are many applications especially related to the environmental, agricultural, and civil engineering.

B. Tillage Effects on Water Flow in Upland Fields

The major purposes of tillage in today's agriculture are, according to Taylor and Ashcroft (9), to prepare a seed bed, to cover seeds during planting operations, to control weeds, and to prepare the land for irrigation. However, careless tillage on a hill slope may cause an acceleration of soil erosion that is not only unfavorable for the farmer but also a loss of natural resources from the viewpoint of global environmental conservation. The most effective way to prevent soil erosion from the beginning is, according to Hillel (10), to avoid the sort of tillage that pulverizes the soil excessively and bares the surface to the action of wind and rain.

Meek et al. (11) investigated the effects of tillage and wheel traffic on the infiltration rate in a furrow-irrigated cotton field. To make these effects comparable, they used a wide tractive research vehicle (WTRV) by which chiseling and traffic treatment could be achieved separately in the same field. The tillage was applied between crops and the wheel traffic was applied during the season to the furrows. Figure 8.7 shows the infiltration rate measured 2 h after flooding in the field. Comparing the effects of tillage on the soil that received the traffic, they concluded that tillage increased the infiltration rate probably due to the reduction in bulk density. Contrarily, comparing the effects of tillage on the soil of no trafficked plots, they found that tillage did not increase and sometimes decreased the infiltration rate due to the disruption of natural channels by tillage.

C. Drainage Efficiency in an Upland Field in Japan

Because of the climate in Japan, which varies from tropical in the south to cool temperate in the north but mostly temperate, the drainage efficiency is a great concern not only for farmers

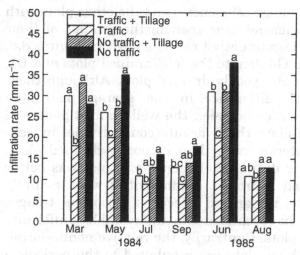

Figure 8.7 Effect of tillage between crops and traffic applied during the season to the furrows on the infiltration rate 2 h after flooding in furrow-irrigated cotton. Values within one date followed by the same letter are not significantly different (Duncan's multiple-range teat, 0.05 level). (After Meek, B. D., Rechel, E. R., Carter, L. M., DeTar, W. R., and Urie, A. L., *Soil Sci. Soc. Am. J. 56*:908–913 (1992). With permission.)

but also for the majority of the citizens. A better drainage of excess rainwater brings about a favorable crop production and a better city life.

Recently, alluvial soil in Fukaya, locating at a basin of Tone River in Saitama prefecture in Japan and reserving a fertile alluvial soil, is indicating a kind of soil degradation and the drainage efficiency is declining. Fukaya is one of the most famous agricultural upland areas, where high-quality Welsh onion is produced. Nakano et al. (12) chose six plots from a 520-ha field in Fukaya, three plots with high drainage efficiencies and the other three with poor drainage efficiencies. The composition of clay minerals in these plots was almost similar and were approximately 20% of Smectite, 30% of Vermiculite, 30 to 40% of Kaolin, and a few percentage of Mica and Chlorite. The most remarkable difference between the plots with good drainage efficiencies and the other plots with poor drainage efficiencies was seen in the profiles of their saturated hydraulic conductivities, as shown in Figure 8.8, where G1, G2, and G3 denote the well-drained plots and B1, B2, and B3 denote the poorly drained plots. Although there was no significant difference in the average saturated hydraulic conductivities between the well-drained plots and the poorly drained plots, the hydraulic conductivities in well-drained plots scattered more than in poorly drained plots. In other words, the drainage efficiency of a plot was better when the saturated hydraulic conductivities were spatially heterogeneous. The integrated effects of periodical tillages and traffics may have caused the unfavorable uniformity of B1, B2, and B3 plots. Contrarily, the relative nonuniformities of the other three plots are attributed to the periodical organic matter applications in the G2 plot, and to the extremely deep plowing that was applied 5 years ago in the G3 plot. The good drainage in the G1 plot may be due to the unknown but presumably careful treatment of the soil by the farmer.

Assuming that the heterogeneity of the saturated hydraulic conductivity in a soil profile is suitable for better drainage efficiency, such improved soil managements such as an effective tillage, an extremely deep tillage, and a nontillage treatment are worthy of more investigations.

III. WATER BALANCE IN FIELDS WITH VEGETATION

It is well known that water balances in soils are influenced considerably by the change in vegetation. Water balance in the root zone of a soil is illustrated in Figure 8.9. Input of water to the root zone is composed of two terms, the precipitation (P) and the upward soil water flux (UF) to the root zone. Output of water from the root zone is composed of three terms: the surface runoff (SR); the ET, which is the sum of evaporation from the soil surface (EV) and transpiration from leaves

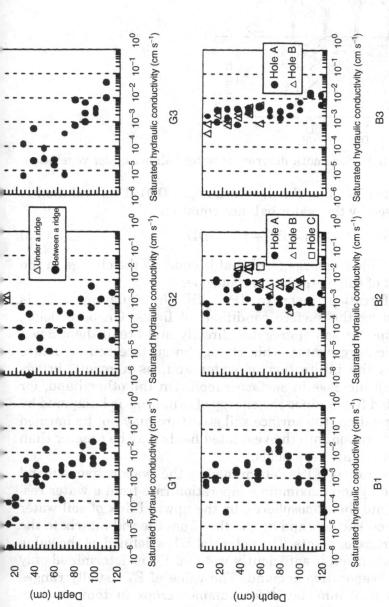

Figure 8.8 Distributions of saturated hydraulic conductivities at well-drained sites G1, G2, and G3, and poorly drained sites B1, B2, and B3 in a Welsh onion production area. In addition to the three samples at each depth (denoted by black circles), additional samples were taken from hole B and hole C at sites B2 and B3. (After Nakano, K., Miyazaki, T., and Nakano, M., *Trans. Jpn. Soc. Irrig. Drain. Reclam. Eng.* 195:101–112 (1998). With permission.)

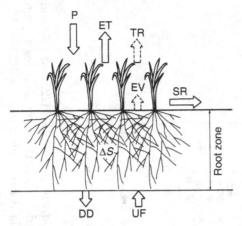

Figure 8.9   Schematic diagram of water balance under vegetation.

of plants (TR); and the deep drainage (DD). Their relation is expressed by the water balance equation

$$\Delta S = (P - SR) - ET + UF - DD \tag{8.3}$$

where ΔS is a storage term that is equal to the change in the amount of soil water in the root zone.

Infiltration, given by (P − SR) in Equation (8.3), is affected by the surface conditions of fields. On bare fields, raindrops hit soil aggregates directly and break them down into microaggregates. This change in surface soil structure impedes the infiltration of water and, as a result, brings about an increase in surface runoff. On the other hand, for vegetated fields, since the energy of raindrops is hampered by the vegetation, the surface soil structure may not be harmed and infiltration into the vegetated fields may be greater than infiltration into bare fields.

The rate of ET is determined by the evaporative demand (defined by the maximum evaporation rate from a water reservoir into the atmosphere), by the upward flux of soil water to the soil surface, and by the flux from the soil matrix to the root surface of plants. The potential ET is defined as the value of ET when soil is sufficiently wet and ET is determined only by the evaporative demand. The value of ET usually ranges from 3 to 6 mm day^{-1} for summer crops in temperature-monsoon area. This value is close to the rate of evaporation from a sufficiently wet bare soil. It should be noted that ET is almost always higher than the rates of evaporation from bare soils, because a plant can collect a great deal of water from the soil of a root zone. This is the reason that water is removed more from soils under vegetation than from soils without plants.

The UF from the deep zone to the root zone increases with a decrease in soil water content in the root zone. The role of this flux is negligibly small for plants under plentiful rainfall but becomes significant or rather crucial when the crops are suffering from drought (13,14).

DD occurs when the water supply exceeds the storage capacity of the soil. However, DD decreases rapidly a few days after a heavy rain and quickly falls below the ET rate; then the flux of soil water turns from downward (DD) to upward (UF).

It should be noted that the ET term has an effect on water balance almost throughout the year. On the other hand, the contribution of the SR and DD terms to water balance are temporal. This means that among the water output terms, the role of ET is particularly important in long-term water balances.

IV. EVAPOTRANSPIRATION

A. Factors Affecting Evapotranspiration

Plants produce sugar by absorbing water from soils, carbon dioxide from the atmosphere through the stomata on the leaves, and energy from solar radiation. This process is called *photosynthesis*. To collect carbon dioxide and solar energy efficiently, plants expand their leaves as far as possible in the atmosphere, which inevitably brings about a huge water loss through the stomata to the atmosphere. Usually, stomata are open only in daytime for photosynthesis, and hence the transpiration occurs during daytime through water vapor diffusion from the stomatal cavity to the ambient air. However, when the humidity of the air is low and the wind blows, plants lose water even in the night. It is striking that almost all of the water extracted by the roots are removed by transpiration, while less than a few percent of the water is used for metabolism and growth. Water-use efficiency, defined as grams of water to produce 1 g of dry matter, is 450 to 950 for C_3 plants such as rice and wheat, and 250 to 350 for C_4 plants such as maize and sorghum. Readers are referred to the book by Kramer (15) for more on the mechanism of transpiration.

Transpiration from plants and evaporation from bare soils are different in both magnitude and duration. In the case of agricultural fields, for example, the soil surface is not completely covered with vegetation, and the contribution of evaporation from soil surface in the ET changes with vegetative growth. In early stages of annual crops, most of the soil surface is exposed to the atmosphere. When the expanded plant leaves shade the soil surface gradually, the ratio of rate of evaporation from soil surface to ET decreases with crop growth. After flowering, ET begins to decrease toward maturity and a part of the soil surface is again exposed by senescence of the leaves, again resulting in an increase in the relative contribution of soil evaporation in ET.

Figure 8.10 shows an example of the rate of pan evaporation, bare soil evaporation, and ET from an upland rice field after a sufficient water supply in the dry season in the tropics (16). In the early stages higher values of transpiration than pan evaporation are due largely to advection because the rice field was isolated in the fallow period. Evaporation from the bare soil surface decreased monotonically with day, indicating the advent of a falling-rate stage (see Chapter 2) starting on the first day. This fall in evaporation rate must be caused by the formation of a dry surface–soil zone due to the intense solar radiation. It is noted in Figure 8.10 that the change in ET value is similar to the change in pan evaporation and solar radiation during the first 6 days. The ET value must be determined by the evaporative demand during this period. The fall in ET value is very quick afterward, probably due to the exhaustion of soil water by plant roots.

Instead of the water balance equation (Equation (8.3)), a number of micrometeorological approaches have been used to determine ET value using energy balance and aerodynamic equations (17,18). These approaches have been applied to ET

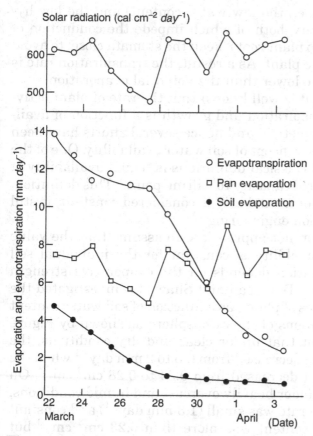

Figure 8.10 Evapotranspiration from an upland rice field after sufficient irrigation. (After Hasegawa, S. and Yoshida, S., *Soil Sci. Plant Nutr.* 28:191–204 (1982). With permission.)

in relatively large areas by irrigation engineers and hydrologists. The combination of the water balance concept, including water uptake by plant roots, as mentioned above, and micrometeorological approach will give more powerful predictions of ET.

B. Soil Water Availability

Transpiration is affected by soil properties such as soil water potential and hydraulic conductivity of the soil. The low soil water potential (i.e., the low water content) and the low hydraulic conductivity, both of which impede the conduction of sufficient water to plant roots, keep the stomata close, to avoid dehydration of the plant. As a result, the transpiration rate is reduced to a value lower than the potential evaporation.

Empirically, it is well known that the rate of plant activities such as transpiration and growth is a function of available soil water depletion, and hence several efforts have been made to define the concept of soil water availability. One of the most famous and classical definitions is "equal availability of soil water from field capacity to wilting point." This definition, in which soil water availability is considered constant, is still useful for irrigation engineering.

It is, however, not appropriate to assume that the value of soil water availability is constant for the individual soil because transpiration depends on the evaporative strength of the atmosphere. Denmead and Shaw (19) investigated the transpiration rates of plants as a function of soil water content for various conditions of the atmosphere as shown by Figure 8.11. It is evident that under clear and dry conditions, the transpiration rate decreased from 6.5 to 2 mm day^{-1} when the soil water content decreased from 0.34 to 0.28 cm^3 cm^{-3}. On the other hand, under heavily overcast and humid conditions, the transpiration rate was small (1.3 mm day^{-1}) and constant when the water content was more than 0.23 cm^3 cm^{-3} but decreased to 0.2 mm day^{-1} when the water content was less than the wilting point (0.22 cm^3 cm^{-3}). It is concluded from Figure 8.11 that available water for plants is not constant but depends on the atmospheric conditions or evaporative demand.

V. SOIL WATER UPTAKE BY PLANT ROOTS

A. Mechanism of Soil Water Uptake

1. Driving Forces of Soil Water Uptake

Water moves in response to gradients in water potential, as discussed in Chapter 2. Since plants absorb water from soils and transpire it from the leaves, it seems that water potential

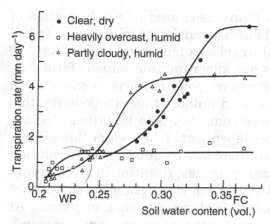

Figure 8.11 Actual transpiration rate as a function of soil water content. WP, wilting point: $-1.5\,\mathrm{kJ\,kg^{-1}}$ soil water content; FC, field capacity. (From Denmead, O. T. and Shaw, R. H., *Agron. J.* 54:385–390 (1962). With permission.)

is lowest at the leaves and highest in the soil in the soil–plant–atmosphere system. This is confirmed by comparing the total water potentials in soils and plants.

The total water potential in soils (ϕ_{soil}) is composed of the matric potential (ϕ_{m}) and the osmotic potential (ϕ_{o}):

$$\phi_{\text{soil}} = \phi_{\text{m}} + \phi_{\text{o}} \tag{8.4}$$

where the values of ϕ_{m} and ϕ_{o} are negative because water is held by soil matrix and dissolves solutes. However, as the concentration of solutes in soils is generally low enough except in saline soils, the effect of ϕ_{o} on the total water potential is usually neglected. Therefore, water in soils moves primarily in response to the gradient of the matric potential. When the soil is wet and the matric potential is relatively high, gravity contributes as a driving force of water and hence gravitational potential (ϕ_{g}) must be added on the right-hand side of Equation (8.4).

The total water potential in plants (ϕ_{plant}) is composed of ϕ_{m}, ϕ_{o}, and pressure potential (ϕ_{p}):

$$\phi_{\text{plant}} = \phi_{\text{m}} + \phi_{\text{o}} + \phi_{\text{p}} \tag{8.5}$$

The matric potential in plants is caused by the hydration of colloidal substances and surface tension at microspaces in cell walls. Osmotic potential in cells ranges from -0.5 kJ kg^{-1} to minus several kilojoules per kilogram; this value is far different from that in soils. Pressure potential acts to expand cell walls as turgor pressure, and unlike other components, this potential is positive. Three components in Equation (8.5) are not independent but interdependent. When water flows into a cell, the osmotic potential increases by dilution of the solution and the pressure potential increases, resulting in more expansive pressure on the cell wall. Each water potential component is not equally important in water movement in plants. At the root surface where water enters roots from surrounding soils, water movement occurs by the osmotic potential gradient. From the roots to the leaves, however, the pressure potential gradient is dominant. Instead of the pressure potential, hydrostatic pressure is often used to express water movement in plants in terms of water movement in vascular vessels: as like flow in a cylindrical tube. In the case of deep-rooted plants or tall trees, water moves a long distance before evaporating from leaves. The difference in elevation from the roots to the leaves must be taken into consideration, and thus the gravitational potential (ϕ_g) is added to the right-hand side of Equation (8.5).

2. Definition of Hydraulic Resistance

Soil water movement obeys Darcy's law, given by

$$q_m = K \frac{\Delta \phi}{\Delta z} \tag{8.6}$$

where q_m is the soil water flux, K is the hydraulic conductivity, and $\Delta \phi$ is the soil water potential difference between two neighboring sites of Δz. If we introduce R as $\Delta z / K$, Equation (8.6) becomes

$$q_m = \frac{\Delta \phi}{R} \tag{8.7}$$

R in this equation is called the hydraulic resistance. This expression came from an analogous equation for the flow of electricity, called Ohm's law, and has been used widely for water movement in the soil and plant systems. van den Honert (20) introduced this idea and proposed the simple expression

$$Q = \frac{\phi_s - \phi_r}{R_s} = \frac{\phi_r - \phi_l}{R_p} = \frac{\phi_s - \phi_l}{R_s + R_p} \tag{8.8}$$

where Q is the flow rate; ϕ_s, ϕ_r, and ϕ_l are water potentials in the soil, the root, and the leaf, respectively; and R_s and R_p are hydraulic resistances in the soil and plant. Since van den Honert presented the idea, many studies have been carried out to evaluate sites and magnitudes of hydraulic resistance in soil–plant systems. Hydraulic resistance is now divided into segments as shown in Figure 8.12, and Equation (8.8) is modified to

$$Q = \frac{\phi_s - \phi_{rs}}{R_s} = \frac{\phi_{rs} - \phi_x}{R_r} = \frac{\phi_x - \phi_{xc}}{R_x}$$

$$= \frac{\phi_{xc} - \phi_{lx}}{R_{stem}} = \frac{\phi_{lx} - \phi_l}{R_{leaf}} \tag{8.9}$$

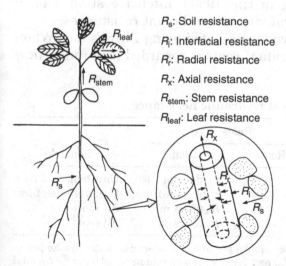

R_s: Soil resistance
R_i: Interfacial resistance
R_r: Radial resistance
R_x: Axial resistance
R_{stem}: Stem resistance
R_{leaf}: Leaf resistance

Figure 8.12 Hydraulic resistances in soil and plant.

in which ϕ_s, ϕ_{rs}, ϕ_x, ϕ_{xc}, ϕ_{lx}, and ϕ_l are water potentials of bulk soil, soil at root surface, root xylem, root crown (where many roots converge into one), leaf xylem, and evaporation site of leaf, respectively. R_s is the soil resistance (rhizosphere resistance), R_r is the resistance between the root surface and xylem (radial or absorption resistance), R_x is the resistance through the root xylem (axial or conductance resistance), R_{stem} is the resistance through the stem, and R_{leaf} is the resistance between the leaf xylem and evaporating site, respectively.

The dimensions of hydraulic resistance in Equation (8.9) depend on the dimensions of the flow rate. Table 8.1 summarizes the hydraulic resistance dimensions of soil and root under different units of flow rate. In this table, the resistances r_r and r'_r act in the radial direction of a root, whereas the resistances R_r and R'_r act in both the radial and axial directions in a set of roots.

3. Factors Affecting Hydraulic Resistances

Equation (8.9) indicates that when a flow rate Q is given, the difference in water potentials between the two adjacent sites in the soil–plant–atmosphere system increases with hydraulic resistance. Beginning in the 1960s, intensive studies have been carried out to evaluate soil and plant resistances.

The soil resistance is evaluated using Darcy's law. When water uptake by individual roots is regarded as a radial flow

Table 8.1 Dimensions of Hydraulic Resistance

Flow rate[a]	Resistance Root	Resistance Soil	Remarks
q (m^2 day^{-1})	r_r	r_s (days m^{-1})	Unit length of root
$q' = qL_v$ (days^{-1})	$r'_r = r_r/L_v$	r'_s (days m)	Unit volume of soil
$Q = qL_T$ (m^3 day^{-1})	R_r	R_s (days m^{-2})	All roots
$Q' = qL_A$ (m day^{-1})	R'_r	R'_s (days)	All roots

[a] q, Water uptake rate; q', water uptake per unit soil volume; Q, water uptake per plant; Q', transpiration rate; L_v, root density per unit volume of soil (m m^{-3}); L_T, total root length (m); L_A, root density per unit ground surface (m m^{-2}).

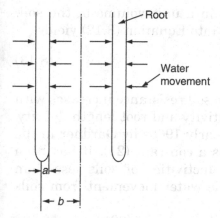

Figure 8.13 Water movement to a single root.

as illustrated by Figure 8.13, water movement under steady-state conditions is expressed as

$$q = 2\pi r K \frac{d\phi_m}{dr} \tag{8.10}$$

where q is the water uptake rate by unit length of root, r is the distance from a root stele, and K is the hydraulic conductivity of the soil. Integrating this equation, the difference between the matric potential of the bulk soil (ϕ_s) and that at the root surface (ϕ_{rs}) is given by

$$\phi_s - \phi_{rs} = \frac{q}{2\pi K} \ln \frac{b}{a} \tag{8.11}$$

where a is the radius of the root and b is the half-distance between adjacent roots. The term b is often expressed as $b = (\pi L_v)^{-1/2}$, where L_v is the root length density (m m^{-3}). As Equation (8.11) focuses on individual roots, this is called a *single-root model*. When water uptake from the unit volume of the soil q' is used for the flow rate, Equation (8.9) becomes

$$q' = qL_v = \frac{\phi_s - \phi_{rs}}{r'_s} \tag{8.12}$$

where r'_s is the soil resistance in a unit volume of the soil. Substitution of Equation (8.11) into Equation (8.12) yields

$$r'_s = \frac{\phi_s - \phi_{rs}}{qL_v} = \frac{1}{2\pi KL_v} \ln\frac{b}{a} \qquad (8.13)$$

This equation indicates that the soil resistance increases with a decrease in hydraulic conductivity and root length density. This idea was proposed in the early 1960s by Gardner in the form $r'_s = (BKL_v)^{-1}$, where B is a constant (21). It has been recognized that hydraulic conductivities of soils less than 10^{-6} cm day^{-1} seriously impede water movement from soils to roots (22–24).

Newman evaluated the soil resistance (22) by assuming that the flow rate in Equation (8.9) is equal to the transpiration rate and by substituting it into Equation (8.11) such that

$$R'_s = \frac{1}{2\pi KL_A} \ln\frac{a}{b} \qquad (8.14)$$

where L_A is the root length per unit ground surface. By quoting the values relating to the parameters in Equation (8.14) from the literature, he calculated the soil resistance R'_s and compared it with the plant resistance presented by Cowan (25). Newman concluded that the plant resistance is higher than the soil resistance when the soil water content is suitable for plant growth, while the soil resistance is higher than the plant resistance when the soil matric potential decreases below −1.5 kJ kg^{-1}. Lawlor (26) and Reicosky and Ritchie (24) supported Newman's conclusion, which is now widely accepted as long as the single-root model is applied.

In the flow path through the plant, roots have the highest hydraulic resistance (27,28) and radial root resistance is believed to be greater than the axial resistance. Even though radial flow pathways from the epidermis to the xylem are not clearly understood (29,30), it is generally accepted that the major resistance to this flow is located at the casparian strip of the endodermis. Taylor and Klepper (31) pointed out that the axial resistance of cotton is negligibly small, whereas that of soybean is appreciable. The axial resistance of dicot plant roots that undergo secondary growth might be smaller than that of monocot plant roots. It is safe to say that the axial resistance is generally small under normal growth conditions (32,33).

The other hydraulic resistance that should be taken into account is the interfacial resistance between the soil and the root surface. Unlike roots in solution culture, roots in soils do not contact water completely. For example, the area of root surface contacting the soil water decreases with soil drying. Roots in large pores whose diameters are larger than the root diameters can contact only a portion of the pore walls. The concept of interfacial resistance is introduced based on these imperfect contacts between soil and roots.

Cowan and Milthorpe (34) first introduced the idea that the surface area of a root contacting soil water is proportional to the degree of saturation. Herkelrath et al. (35) showed that the water uptake rate obtained by a single-root model over-estimated the rate measured by experiment, and they introduced the contact model into the single-root model to explain their results. There is, however, little direct evidence that imperfect contact impedes water movement (36).

Both roots and soils shrink by drying. Huck et al. (37) observed that a cotton root grown in a rhizotron shrank to 60% of its maximum diameter at midday on a fine day. Passioura (33) pointed out that if the main resistance of radial flow through the root exists at endodermis, water potential in the cortex should be close to that in the soil and the root might not shrink under normal field conditions, where water uptake rates are usually lower than those obtained in laboratory experiments.

B. Equation of Soil Water Uptake

The equation of soil water movement in a root zone is given by

$$\frac{\partial \theta}{\partial t} = \frac{\partial}{\partial z}\left[K(\theta)\left(\frac{\partial \psi_m}{\partial z} - 1\right)\right] - s \tag{8.15}$$

where θ is the volumetric water content; t is the time; z is the depth; K is the hydraulic conductivity, which is a function of θ; ψ_m ($= \phi_m/g$) is the matric head of soil water; and s is the extraction by roots. Since the unsaturated hydraulic conductivity $K(\theta)$ generally decreases radically with decreased soil water content, the extraction term s becomes predominant with decreased soil water content in Equation (8.15) (38).

Many researchers have proposed models of the extraction term s as reviewed by Molz (30). For example, Molz and Remson (39) proposed an empirical model

$$s = \frac{-1.6T}{v^2}z + \frac{1.8T}{v} \tag{8.16}$$

where T is the transpiration rate and v is the depth of the root zone. If the root zone is divided into four layers of equal length from the top, this equation indicates that water consumed by each layer becomes 40, 30, 20, and 10%, respectively, of the transpiration rate. Hillel et al. (40) used a potential model

$$s = \frac{\phi_s - \phi_{xo}}{r'_s + r_{ra}} \tag{8.17}$$

where ϕ_s is the water potential of the soil; ϕ_{xo} is the water potential of the root crown; r_{ra} is the root resistance, which is the sum of the radial and axial resistances; and r'_s is the soil resistance equal to $(BKL_v)^{-1}$ as given in the preceding section. In this model, however, the difference in flow resistances due to the portion of the root xylem was not taken into account.

Feddes et al. (41) proposed another water uptake model in the form

$$s = \begin{cases} 0 & 0 \le \theta \le \theta_w \\ s_{max}\frac{\theta - \theta_w}{\theta_d - \theta_w} & \theta_w \le \theta \le \theta_d \\ s_{max} & \theta_d \le \theta \le \theta_{an} \\ 0 & \theta_{an} \le \theta \le \theta_s \end{cases} \tag{8.18}$$

where s_{max} is the maximum rate of water extraction by plant roots, θ_w is the soil water content at the wilting point, θ_d is the lowest value of θ at which $s = s_{max}$, θ_{an} is the highest value of θ at which $s = s_{max}$, and θ_s is the saturation water content. This model takes account of the explicit effects of soil water content on water uptake.

When a root zone is divided into many slices, as shown in Figure 8.14, the transpiration rate is expressed as the sum of the extraction rate,

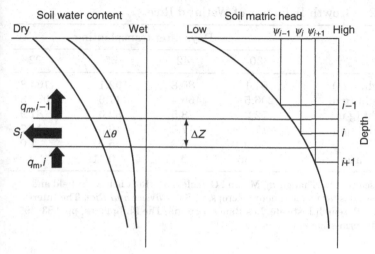

Figure 8.14   Water balance in a root zone.

$$T = \sum_{i=1}^{n} s_i \tag{8.19}$$

where n is the number of slices contributing to the water uptake. When the summation of Equation (8.19) exceeds the evaporative demand (denoted by T_{max}) in the calculation, the summation is truncated up to the rth layer so as to satisfy

$$T_{max} = \sum_{i=1}^{r} s_i \quad (r \leq n) \tag{8.20}$$

The small value of r means that roots prefer to extract water from shallow soil layers. The larger the axial resistance through the root xylems, the smaller may be the value of r (42).

C.   Distribution of Crop Roots

Table 8.2 shows root and top growth values of rice under flooded conditions (43). The root length reached a maximum 30 days after transplantation, although the root weight was still increasing at head development, at 65 days. The plant heights increased until ripening, at 92 days. The root–shoot

Table 8.2 Growth Behavior of Wetland Rice

	Days after transplanting			
	30	42	65	92
Plant height (cm)	64.1	80.8	102.1	104.8
Shoot weight (g m^{-2})	206.5	465.4	747.0	1079.3
Root weight (g m^{-2})	72.2	78.5	82.0	81.9
Root length (km m^{-2})	17.7	17.5	16.5	14.7
Leaf area index	3.0	6.3	5.0	1.0
Root shoot ratio	0.35	0.17	0.11	0.08

Source: Hasegawa, S., Thangaraj, M., and O'Toole, J.C., Root behavior: field and laboratory studies for rice and nonrice crops, in Soil Physics and Rice, The International Rice Research Institute, Los Banos, Laguna, The Philippines, pp. 383–395 (1985). With permission.

ratio measured 30 days after transplantation was 35%, and it decreased to 8% at maturity. In other words, the development of the roots preceded growth of the top.

Root length density is defined by either the root length per unit volume of a soil (L_v: m m^{-3}) or the root length per unit surface of a soil (L_a: m m^{-2}). The former is used to determine the root length distribution with depth, and the latter is used to determine the amount of roots per unit ground surface or crack surface. Figure 8.15 shows root length distributions of three upland crops and wetland rice in the Philippines (43,44). The crops were sown or transplanted in the dry season and grown under a sufficient water supply. As roots grow not only vertically but also horizontally, roots were taken at zero, one fourth, and one half distance between rows by a core sample method at flowering times. Mean values of root length densities at these sites were used in Figure 8.15. The root length density of wetland rice was extremely high in the top 10 cm, and the root length density of sorghum was lowest throughout the layers. Huge amounts of the roots of rice are due to its tillering character, which is also applicable to wheat. The depth at which 90% of total root length is contained is 20 cm for wetland rice, 40 cm for wheat, 50 cm for sorghum, and 60 cm for maize.

Root development is influenced by the water supply. Figure 8.16 shows root distributions of the same variety of

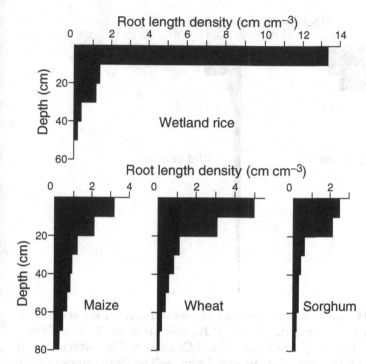

Figure 8.15 Root length density of wetland rice and three upland crops. (After Hasegawa, S., Thangaraj, M., and O'Toole, J.C., Root behavior: field and laboratory studies for rice and nonrice crops, in *Soil Physics and Rice,* The International Rice Research Institute, Los Banos, Laguna, The Philippines, pp. 383–395 (1985); Hasegawa, S., Parao, F. R., and Yoshida, S., *Root Development and Water Uptake Under Field Condition,* The International Rice Research Institute, Los Banos, Laguna, The Philippines (1979). With permission.)

upland rice grown in wet and dry seasons. Irrigation by sprinkler at 5- to 7-day intervals was used on dry-season rice, whereas frequent rains were the only source of water for wet-season rice. Root length densities of the dry-season rice were higher than those of the wet-season rice except for the surface 5 cm. Frequent, limited water applications wet the soil to a shallow depth and stimulate the development of roots at this depth. Water management therefore alters the distribution of roots considerably (45).

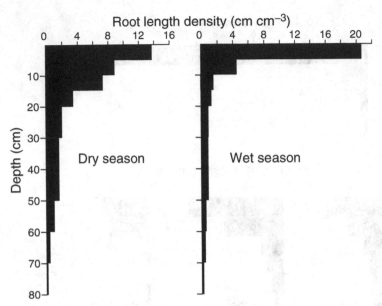

Figure 8.16 Root length density of upland rice at different seasons. (After Hasegawa, S., Parao, F. R., and Yoshida, S., *Root Development and Water Uptake Under Field Condition*, The International Rice Research Institute, Los Banos, Laguna, The Philippines (1979). With permission.)

Root growth is strongly influenced by the existence of macropores. Figure 8.17 shows an example of soybean roots grown in a loam soil and a clay soil. The loam soil was used for upland crop cultivation, whereas the clay soil had been used as a wetland rice culture and then converted to upland use. This conversion caused drying cracks to develop in the subsoil. Two soil blocks of 10×15 cm^2 surface area were taken from each soil. In the clay subsoil, roots were developed in the soil matrix and on the crack surfaces. Although the crack surface area in the soil blocks was different at individual depths, about half the roots were distributed on the crack surfaces in the subsoil in this case. But the total root length of the clay soil was almost the same as that of the loam soil. This figure demonstrates the importance of roots grown in cracks on water uptake.

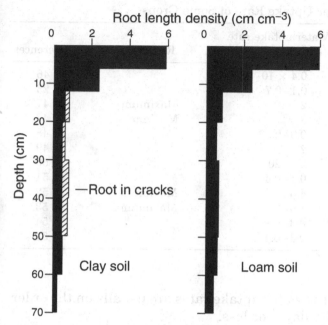

Figure 8.17 Root length density of soybean grown in different soils.

D. Water Uptake Rate and Water Extraction Pattern

1. Water Uptake Rate

The water uptake rate is obtained experimentally by dividing the amount of water absorbed by roots in a given volume of soil by the length of roots in it. Water uptake rates of individual roots are not the same but differ depending on the age and distance from a root tip. Water is absorbed more in the mature zone than in differentiated and old zones. However, Kramer (15) has pointed out that the major percentage of the water and salt absorption by perennial plants probably occurs through old suberized roots. In most studies dealing with a root system, the water uptake rate is assumed to be the same regardless of age. Table 8.3 shows water uptake rates for some crops. Even though experimental conditions such as climate

Table 8.3 Water Uptake Rate of Some Crops

Crop	Water uptake rate $(cm^3\ cm^{-1}\ day^{-1})$	Remarks	References
Cotton	0.4×10^{-2}		46
	0.1–0.7		23
	2	Maximum	47
Maize	3	Maximum	47
	0.3–24.3		48
	2		49
Soybean	2.5–20		50
	0.1–0.4		51
	4.6	Maximum	52
	1.0	Maximum	53
	0.3–0.9		54
Sorghum	0.3–0.4		54

and soil differ, the water uptake rates are usually on the order of $0.01\ cm^3\ cm^{-1}\ day^{-1}$ or less.

2. Water Extraction Pattern

When a soil profile is thoroughly wet, plants extract most soil water from shallow, dense rooted layers. With the decrease of soil water content in the shallow layers, water retained in deeper layers begins to make a larger contribution to transpiration. Figure 8.18 shows the water depletion patterns of upland rice after sufficient irrigation on March 21. The depletion rate is equal to the change in water content per day and expressed as millimeters per day. The ET rate and root length density of this crop are shown in Figure 8.10 and Figure 8.16 (left). Most of soil water was extracted from the surface 30 cm until 3 days after the irrigation (Figure 8.18, left). As the soil water in shallow layers decreased, the major root zone contributing water uptake was shifted to 30 to 50 cm (Figure 8.18, middle). Finally, most of the soil water in the upper layers had been consumed and the plant was unable to extract sufficient water to satisfy the evaporative demand because of sparse roots in deep layers even though the soil in the deepest layer contained enough water (Figure 8.18, right).

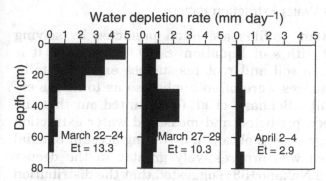

Figure 8.18 Water extraction patterns after the irrigation on March 21. Et, evapotranspiration rate, mm day^{-1}. (After Hasegawa, S. and Yoshida, S., *Soil Sci. Plant Nutr.* 28:191–204 (1982). With permission.)

The reason the plant extracted water preferentially from shallow layers was attributed to the high root densities in these layers. However, the relative root density did not coincide with water depletion patterns, even though the transpiration rate was governed by the evaporative demand (Table 8.4). One reason for this water extraction pattern is that water uptake rates in shallow layers were lower due to the aged and suberized portion of roots. The other reason that might be taken into account is the existence of axial resistance.

Table 8.4 Relative Water Consumption and Relative Root Density at Different Soil Depths

Depth (cm)	Relative water consumption (%)		Relative root length density (%)
	March 22 to 28	March 28 to April 4	
0–20	43	18	70
20–40	27	34	15
40–60	21	29	12
60–80	10	19	3

Source: Hasegawa, S. and Yoshida, S., *Soil Sci. Plant Nutr.* 28:191–204 (1982). With permission.

3. Prediction of Water Extraction Pattern

Soil water content profiles are often predicted by solving Equation (8.15) with s of Equation (8.17). In practice, it is not easy to obtain soil and root resistances experimentally, and these resistances were often applied so as to fit the experimental results. Belmans et al. (55) pointed out that the better fit between predicted and measured water extraction patterns for ryegrass was obtained by taking account of axial resistance, which was progressively greater in the deeper roots. Ishida and Nakano (53) suggested that the distribution of roots and the rate of ET affected the early stage of the water extraction pattern after watering, whereas axial resistance affected the water extraction pattern considerably throughout all extraction stages. Hillel and Talpaz (56) discussed the effects of root growth on the water extraction pattern, which was usually ignored in the calculation. They showed that the root extension, which enlarged rooting depth, influenced the extraction pattern more significantly than did root proliferation, which increased root length within the rooting depth.

Here we try to predict the water extraction pattern by use of Equations (8.18) to (8.20) as the extraction term. As described previously, this extraction term has two assumptions. The first assumption is that the water uptake rate is expressed as a function of soil water content as shown in Equation (8.18). The second assumption is that the resistance of water flow in root xylem is appreciable, as shown in Equations (8.19) and (8.20). A model introducing the two assumptions mentioned was applied to predict successive soil water content profiles by an upland rice whose ET rate and root length density are shown in Figure 8.10 and Figure 8.16 (left). In the calculation we neglected water movement driven by soil matric potential, which means that the water depletion rate is equal to the product of water uptake rate and root length density. Table 8.5 shows the calculated values of water uptake rates q_0 and root length density in individual layers. This table suggests that the older surface roots have a lower uptake rate than the deeper young roots. Successive soil water content profiles by the model are shown in Figure 8.19.

Table 8.5 Root Density and Water Uptake at Field Capacity by the Model

Depth (cm)	Root length density L_v (cm cm^{-3})	Uptake rate q_o (cm^3 cm^{-1} day^{-1})
0–10	11.22	0.45×10^{-2}
10–20	5.22	0.52
20–30	1.87	1.05
30–40	1.62	2.42
40–50	1.62	2.00
50–60	1.21	0.93
60–70	0.63	1.48
70–80	0.23	2.87

Source: Hasegawa, S., *Soil Phys. Cond. Plant Growth Jpn. 44*:14–22 (1981). With permission.

The agreement of successive soil water contents between the measured and predicted values are good at 0 to 30 and 70 to 80 cm, but the predicted values were overestimated at

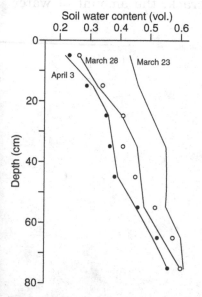

Figure 8.19 Successive soil water content profiles by the model. Open and solid circles show values obtained from the model on March 28 and April 3.

50 to 70 cm. The calculated potential transpiration rate ceased on the fourth day, 2 days before the experiment. However, the total amount of water extracted by the model was 109.0 mm, which agreed fairly well with the experimental result of 108.2 mm.

E. Water Uptake by Roots in Cracks

1. Estimation of Daily Water Uptake

Most models of water uptake assume that soil is homogeneous and that roots create their own pores uniformly. However, plant roots grow preferentially along macropores such as drying cracks and structural bed surfaces. To estimate water uptake by roots developed in drying cracks, Hasegawa and Sato (57) assumed that vertical cracks penetrating the subsoil divide a soil into rectangular prisms as shown in Figure 8.20. In this model, plant roots growing into cracks contact one or both sides of the crack walls, depending on the width of the cracks and the diameter of the roots. When roots are in contact with both sides of a crack, the amount of water

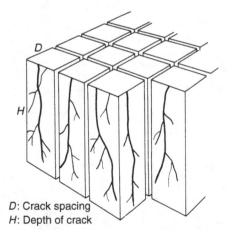

D: Crack spacing
H: Depth of crack

Figure 8.20 Crack model. (After Hasegawa, S. and Sato, T., *Soil Sci. 143*:381–386 (1987). With permission.)

extracted per day from one side of the crack wall, Q', is given by

$$Q' = qL_a H \frac{D}{2} \tag{8.21}$$

where q is the water uptake rate by a unit length of root (cm^3 cm^{-1} day^{-1}), L_a is the root length density defined by the length of roots per unit surface area in the cracks (cm cm^{-2}), H is the crack depth, and D is the separation of adjacent cracks. If the width of a crack is greater than the root diameter, half the roots in a crack will contact one side of the crack wall, and the amount of water extracted from one side of the crack wall will also be given by Equation (8.21). Because one rectangular prism with upper surface area D^2 has four vertical crack walls, the daily water uptake Q (cm day^{-1}) from a unit ground surface is given by

$$Q = 4Q'D^{-2} \tag{8.22}$$

Substitution of Equation (8.21) into Equation (8.22) yields

$$Q = 2qL_a HD^{-1} \tag{8.23}$$

in which all the parameters except q are obtained from field surveys.

This model was applied to a soybean field of heavy clay converted from a wetland rice field. The mean separation and depth of the drying cracks were 15 and 45 cm, respectively. Almost all of the soybean roots were not observed in the soil matrix but developed along crack walls. Figure 8.21 shows root length densities on each crack wall surveyed, whose values ranged from 0 to 2.74 cm cm^{-2}. The mean value of the root density obtained from the six cracks was 0.92 cm cm^{-2}. The water uptake rates q were assumed to be 0.01 or 0.02 cm^3 cm^{-1} day^{-1}, depending on Refs. (50, 51, 54, 58).

Substituting these values into Equation (8.23), the daily water uptake per unit ground surface was estimated to be 0.7 and 1.3 mm day^{-1}, corresponding to 20 and 40% of the daily ET rates measured in the field. Since these amounts of water uptake by roots in cracks are at least one order higher than

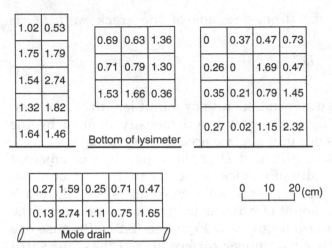

Figure 8.21   Root length density in cracks. (After Hasegawa, S. and Sato, T., *Soil Sci. 143*:381–386 (1987). With permission.)

the upward soil water flux from the subsoil to the surface soil, the decrease in subsoil water content is attributed exclusively to water uptake by roots in the cracks.

2.   Soil Water Content Profiles Toward Cracks with Roots

Ritchie and Adams (59) recognized that water content in a soil matrix decreased toward a crack that was open to the atmosphere. If plant roots spread along a crack wall, the water content in the soil matrix will be lower toward the crack wall, due to additional water uptake by the plant.

Sato et al. (60) solved the equation of water movement in a soil matrix toward a crack wall and compared the solution with experimental data. Horizontal water movement in a soil matrix toward a vertical crack wall is expressed by the one-dimensional flow equation

$$\frac{\partial \theta}{\partial t} = \frac{\partial}{\partial x}\left[K(\psi_m)\frac{\partial \psi_m}{\partial x} \right] \tag{8.24}$$

where θ is the soil water content, x is the distance from the crack wall, and K is the unsaturated hydraulic conductivity, which is a function of the matric head ψ_m. Water extraction per unit crack wall area is given by

$$S = q\frac{L_a}{2} \tag{8.25}$$

based on the same idea applied to Equation (8.21). Soil water content distributions toward a crack wall were solved under the following initial and boundary conditions:

$$\theta = \theta_i, \quad 0 \le x \le D/2, \quad t = 0 \tag{8.26}$$

$$K(\psi_m)\frac{\partial \psi_m}{\partial x} = S, \quad x = 0, \quad t \ge 0 \tag{8.27}$$

$$\frac{\partial \psi_m}{\partial x} = 0, \quad x = D/2, \quad t \ge 0 \tag{8.28}$$

where θ_i is the initial soil water content and D is the separation of adjacent cracks. Equation (8.28) means that soil water does not move at the middle of a soil block put between two adjacent cracks. The equation was solved under the condition that roots extract soil water during the day but not at night.

Figure 8.22 shows two examples of soil water profiles solved by under conditions of 7-day water uptake, a D value

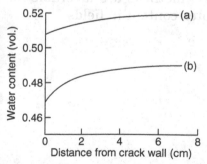

Figure 8.22 Calculated soil water content gradients toward a crack with roots after 7 days of extraction. Initial water content of (a) 0.52 and (b) 0.49. (After Sato, T., Hasegawa, S., Nakano, M., and Miyazaki, T., *Soil Phys. Cond. Plant Growth* 60:24–27 (1990). With permission.)

of 15 cm, a value of q of 0.01 cm^3 cm^{-1} day^{-1}, and a root density L_a of 1.0 cm cm^{-2}. The decrease in water content toward the crack wall was smaller in a soil whose initial water content was large (a) than in a soil whose initial water content was small (b).

On the other hand, the measured water content profile did not necessarily indicate a decrease toward the crack wall as shown in Figure 8.23, where the crack was covered with soybean roots whose root density L_a was 1.5 cm cm^{-2}. The nonuniformity of the soil matrix around the soybean roots may be the reasons for the scattering of the measured soil water profile.

VI. EFFECTS OF VEGETATION ON SOIL WATER REGIME

Typical differences in vegetation in fields are observed, for example, among forest, pasture, and wheat fields. One of the simplest ways to grasp the effects of these differences on a soil water regime is to measure soil water profiles simultaneously and continuously at locations where the environmental conditions are similar except for vegetation. The heat probe and TDR (time–domain reflectometer) methods are favorable for continuous monitoring of soil water content in fields.

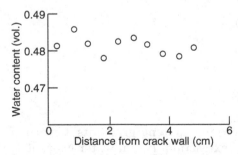

Figure 8.23 Soil water contents toward a crack with roots in the field. (After Sato, T., Hasegawa, S., Nakano, M., and Miyazaki, T., *Soil Phys. Cond. Plant Growth* 60:24–27 (1990). With permission.)

Kasubuchi (61) measured the thermal conductivities of two different soils of various water contents and expressed them as functions of volumetric water content as shown in Figure 8.24, where all the thermal conductivities increased with water content. The thermal conductivities of soils were measured precisely in a few minutes using the twin transient-state cylindrical-probe method (twin heat probe method) (62). The thermal conductivity of silty clay loam was higher than that of volcanic ash soil because the solid content of the former (0.521 cm^3 cm^{-3}) was larger than that of the latter (0.201 cm^3 cm^{-3}). The values of thermal conductivities depended somewhat on temperature, which in practice may be neglected in field measurements.

Kasubuchi (63) applied this heat probe method to monitoring temperature and soil moisture in an area, shown in Figure 8.25, where forest, pasture, and wheat fields were situated in order from south to north. This area was located at 43° north latitude and 141° east longitude within 1 km, and

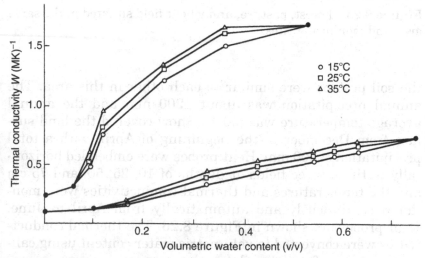

Figure 8.24 Thermal conductivities of silty clay loam (upper three curves) and volcanic ash soil (lower three curves) versus volumetric water content and their dependence on temperature. (After Kasubuchi, T., *Bull. Natl. Inst. Agric. Sci. Ser. B 33*:1–54 (1982). With permission.)

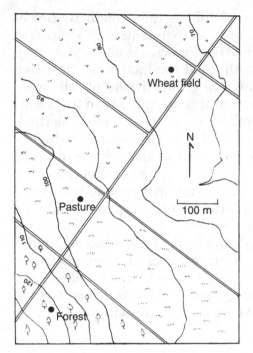

Figure 8.25   Forest, pasture, and wheat field situated in the same area and monitoring sites.

the soil profiles were similar to each other in this area. The annual precipitation was about 1200 mm and the annual average temperature was 8.0°C. Snow covered the land surface from December to the beginning of April, with a total precipitation of 400 mm. Heat probes were embedded horizontally in these three fields at depths of 10, 25, 50, and 75 cm and the temperatures and thermal conductivities were monitored continuously and automatically from April to June. Heat probes are shown in Figure 8.26. The thermal conductivities were converted to volumetric water content using calibration curves for each soil.

Figure 8.27 shows the changes in water content in these fields over a 3-month period. Water content in the forest was almost always lower than it was in the other two fields at every depth, due partly to the low groundwater level and

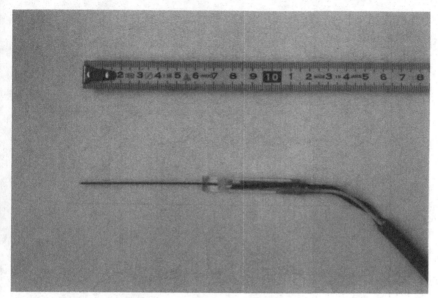

Figure 8.26   Heat probe.

partly to the relatively even absorption of soil water at every depth by well-developed roots in woods and plants in forest. A remarkable decrease in water content during the 3-month period was observed at depths of 10 and 25 cm in the wheat field. This feature was logically attributed to the rapid growth of wheat during this season accompanied by a rapid increase in water uptake by wheat roots. On the other hand, the water content at depths of 50 and 75 cm in the wheat field was high and constant during this period, due presumably to the high groundwater level in the wheat field. Although the water content in the pasture fluctuated at every depth following changes in atmospheric conditions, the decrease in soil water was rather small at all depths. This feature indicates that even if the soil water content decreased by water uptake by grass, the deficiency was compensated quickly by upward water flux from the subsoil.

Figure 8.28 shows changes of soil temperature in forest, pasture, and wheat field. Since snow remained longer in the forest than in the other two fields in early April, the surface

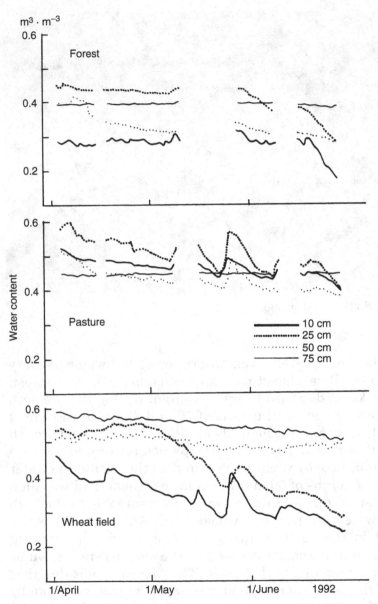

Figure 8.27 Changes in soil water content with different vegetation.

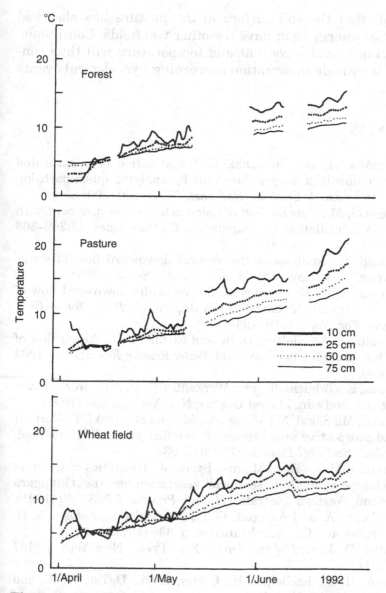

Figure 8.28 Changes in soil temperature with different vegetation.

temperature of the forest was very low during this period. The soil temperature in the pasture increased more than in the other two fields during this 3-month period. This feature indicates that the soil surface in the pasture has absorbed more solar energy than have the other two fields. Continuous monitoring of water content and temperature will thus contribute to provide information concerning hydrological events in fields.

REFERENCES

1. Hamada, H. and Miyazaki, T., Investigation of unsaturated water flow in a deep vodose zone by applying quasi-partition law to ^{222}Rn, *J. Environ. Radioact.* 71:89–100 (2004).
2. Noguchi, M., New method of radon activity measurement with liquid scintillation (in Japanese), *Radioisotopes* 13:362–366 (1964).
3. Takagi, S., Analysis of the vertical downward flow of water through a two-layered soil, *Soil Sci.* 90:98-1–3 (1960).
4. Yamazaki, F., Researches on the vertically downward flow of water through layered soil (in Japanese), *Bull. Tokyo Coll. Agric. For. No. 1*:1–19 (1948).
5. Srinilta, S. A., Nielsen, D. R., and Kirkham, D., Steady flow of water through a two-layer soil, *Water Resour. Res.* 5:1053–1063 (1969).
6. Iwata, S., Tabuchi, T., and Warkentin B. P., *Soil–Water Interactions*, 2nd edn, Marcel Dekker, New York, p. 290 (1995).
7. Khalil, M., Sakai, M., Mizoguchi, M., and Miyazaki, T., Current and prospective applications of zero flux plane (ZFP) method, *J. Jpn. Soc. Soil Phys.* 95:75–90 (2003).
8. Sharma, M. L., Bari, M., and Byrne, J., Dynamics of seasonal recharge beneath a semiarid vegetation on the Gnangara Mound, Western Australia, *Hydrol. Process.* 5:383–389 (1991).
9. Taylor, S. A. and Ashcroft, G. L., *Physical Edaphology*, W. H. Freeman and Co., San Francisco, p. 334 (1972).
10. Hillel, D. J., *Out of the Earth*, Free Press, New York, p. 167 (1991).
11. Meek, B. D., Rechel, E. R., Carter, L. M., DeTar, W. R., and Urie, A. L., Infiltration rate of a sandy loam soil: effects of traffic, tillage, and plant roots, *Soil Sci. Soc. Am. J.* 56:908–913 (1992).
12. Nakano, K., Miyazaki, T., and Nakano, M., Physical fertility of soils and the drainage efficiencies of upland field, *Trans. Jpn. Soc. Irrig. Drain. Reclam. Eng.* 195:101–112 (1998).

13. Van Bavel, C. H. M., Brust, K. J., and Stirk, G. B., Hydraulic properties of a clay loam soil and the field measurement of water uptake by roots. II. The water balance of the root zone, *Soil Sci. Soc. Am. Proc. 32*:317–321 (1968).

14. Reicosky, D. C., Doty, C. W., and Campbell, R. B., Evapotranspiration and soil water movement beneath the root zone of irrigated and nonirrigated millet (*Panicum miliaceum*), *Soil Sci. 124*:95–101 (1977).

15. Kramer, P. J., *Plant and Soil Water Relationships: A Modern Synthesis*, McGraw-Hill, New York (1969).

16. Hasegawa, S. and Yoshida, S., Water uptake by dryland rice root system during soil drying cycle, *Soil Sci. Plant Nutr. 28*:191–204 (1982).

17. Monteith, J. L., The development and extension of Penman's evaporation formula, in *Applications of Soil Physics*, D. Hilell, Ed., Academic Press, New York, pp. 247–253 (1980).

18. Jones, F. E., *Evaporation of Water: With Emphasis on Applications and Measurements*, Lewis Publishers, Michigan (1992).

19. Denmead, O. T. and Shaw, R. H., Availability of soil water to plants as affected by soil moisture content and meteorological conditions, *Agron. J. 54*:385–390 (1962).

20. van den Honert, T. H., Water transport as a catenary process, *Discuss. Faraday Soc. 3*:146–153 (1948).

21. Gardner, W. R., Relation of root distribution to water uptake and availability, *Agron. J. 56*:41–45 (1964).

22. Newman, E. I., Resistance to water flow in soil and plant. I. Soil resistance in relation to amounts of root: theoretical estimates, *J. Appl. Ecol. 6*:1–12 (1969).

23. Taylor, H. M. and Klepper, B., Water uptake by cotton root systems: an examination of assumptions in the single root model, *Soil Sci. 120*:57–67 (1975).

24. Reicosky, D. C. and Ritchie, J. T., Relative importance of soil resistance and plant resistance in root water absorption, *Soil Sci. Soc. Am. J. 40*:293–297 (1976).

25. Cowan, I. R., Transport of water in the soil–plant–atmosphere system, *J. Appl. Ecol. 2*:221–239 (1965).

26. Lawlor, D. W., Growth and water use of *Lolium Perenne*. I. Water transport, *J. Appl. Ecol. 9*:79–98 (1972).

27. Jensen, R. D., Taylor, S. A., and Wiebe, H. H., Negative transport and resistance to water flow through plants, *Plant Physiol. 36*:633–638 (1961).

28. Neumann, H. H., Thurtell, G. W., and Stevenson, K. R., *In situ* measurement of leaf water potential and resistance to water flow in corn, soybean, and sunflower at several transpiration rates, *Can. J. Plant Sci. 54*:175–184 (1974).

29. Newman, E. I., Root and soil water relations, in *The Plant Root and Its Environment*, E. W. Carson, Ed., University of Virginia, Charlottesville, VA, pp. 363–440 (1974).

30. Molz, F. J., Simulation of plant–water uptake, in *Modeling Wastewater Renovation by Land Application*,I. K. Iskandar, Ed., Wiley, New York, pp. 69–91 (1981).

31. Taylor, H. M. and Klepper, B., The role of rooting characteristics in the supply of water to plants, *Adv. Agron. 30*:99–128 (1978).

32. Klepper, B., Managing root systems for efficient water use: axial resistances to flow in root systems — anatomical considerations, in *Limitations to Efficient Water Use in Crop Production*,H. M. Taylor, W. R. Jordan, and T. R. Sinclair, Eds., ASA–CSSA–SSSA, Madison, pp. 115–125 (1983).

33. Passioura, J. B., Water transport in and to roots, *Annu. Rev. Plant Physiol. Plant Mol. Biol. 39*:245–265 (1988).

34. Cowan, I. R. and Milthorpe, F. L., Plant factors influencing the water status of plant tissues, in *Water Deficits and Plant Growth*,T. T. Kozlowski, Ed., Academic Press, New York, pp. 137–193 (1968).

35. Herkelrath, W. N., Miller, E. E., and Gardner, W. R., Water uptake by plants: II. The root contact model, *Soil Sci. Soc. Am. J. 41*:1039–1043 (1977).

36. Faiz, S. M. A. and Weatherley, P. E., Root contraction in transpiring plants, *New Phytol. 92*:333–343 (1982).

37. Huck, M. G., Klepper, B., and Taylor, H. M., Diurnal variations in root diameter, *Plant Physiol. 45*:529–530 (1970).

38. Ritchie, J. T., Burnett, E., and Henderson, R. C., Dryland evaporative flux in a subhumid climate. III. Soil water influence, *Agron. J. 64*:168–173 (1972).

39. Molz, F. J. and Remson, I., Extraction term models of soil moisture use by transpiring plants, *Water Resour. Res. 6*:1346–1356 (1970).

40. Hillel, D., Talpaz, H., and van Keulen, H., A macroscopic-scale model of water uptake by a nonuniform root system and of water and salt movement in the soil profile, *Soil Sci. 121*:242–255 (1976).

41. Feddes, R. A., Kowalik, P., Malinka, K. K., and Zaradny, H., Simulation of field water uptake by plants using a soil water dependent root extraction function, *J. Hydrol. 31*:13–26 (1976).
42. Hasegawa, S., A simplified model of water uptake by dryland rice root system (in Japanese), *Soil Phys. Cond. Plant Growth Jpn. 44*:14–22 (1981).
43. Hasegawa, S., Thangaraj, M., and O'Toole, J. C., Root behavior: field and laboratory studies for rice and nonrice crops, in *Soil Physics and Rice,* The International Rice Research Institute, Los Banos, Laguna, The Philippines, pp. 383–395 (1985).
44. Hasegawa, S., Parao, F. R., and Yoshida, S., *Root Development and Water Uptake Under Field Condition,* The International Rice Research Institute, Los Banos, Laguna, The Philippines (1979).
45. Klepper, B., Taylor, H. M., Huck, M. G., and Fiscus, E. L., Water relations and growth of cotton in drying soil, *Agron. J. 65*:307–310 (1973).
46. Lang, A. R. G. and Gardner, W. R., Limitation to water flux from soils to plants, *Agron. J. 62*:693–695 (1970).
47. Bar-Yosef, B. and Lambert, J. R., Corn and cotton growth in response to soil impedance and water potential, *Soil Sci. Soc. Am. J. 45*:930–935 (1981).
48. Allmaras, R. R., Nelson, W. W., and Voorhees, W. B., Soybean and corn rooting in southwestern Minnesota. II. Root distributions and related water inflow, *Soil Sci. Soc. Am. Proc. 39*:771–777 (1975).
49. So, H. B., Aylmore, L. A. G., and Quirk, J. P., Measurement of water fluxes and potentials in a single root–soil system. I. The tensiometer–photometer system, *Plant Soil 45*:577–594 (1976).
50. Willatt, S. T. and Taylor, H. M., Water uptake by soya-bean roots as affected by their depth and by soil water content, *J. Agric. Sci. Camb. 90*:205–213 (1978).
51. Eavis, B. W. and Taylor, H. M., Transpiration of soybeans as related to leaf area, root length, and soil water content, *Agron. J. 71*:441–445 (1979).
52. Blizzard, W. E. and Boyer, J. S., Comparative resistance of the soil and the plant to water transport, *Plant Physiol. 66*:809–814 (1980).
53. Ishida, T. and Nakano, M., Soil water movement in the soil–plant–atmosphere continuum (in Japanese), *Trans. Jpn. Soc. Irrig. Drain. Reclam. Eng. 92*:26–34 (1981).

54. Burch, G. J., Smith, R. C. G., and Mason, W. K., Agronomic and physiological responses of soybean and sorghum crops to water deficits. II. Crop evaporation, soil water depletion, and root distribution, *Aust. J. Plant Physiol. 5*:169–177 (1978).

55. Belmans, C., Feyen, J., and Hillel, D., An attempt at experimental validation of macroscopic-scale models of soil moisture extraction by roots, *Soil Sci. 127*:174–186 (1979).

56. Hillel, D. and Talpaz, H., Simulation of root growth and its effect on the pattern of soil water uptake by a nonuniform root system, *Soil Sci. 121*:307–312 (1976).

57. Hasegawa, S. and Sato, T., Water uptake by roots in cracks and water movement in clayey subsoil, *Soil Sci. 143*:381–386 (1987).

58. Hasegawa, S. and Sato, T., Soil water movement in the vicinity of soybean roots determined by root plane experiment, *Trans. Jpn. Soc. Irrig. Drain. Reclam. Eng. 117*:17–24 (1985).

59. Ritchie, J. T. and Adams, J. E., Field measurement of evaporation from soil shrinkage cracks, *Soil Sci. Soc. Am. Proc. 38*:131–134 (1974).

60. Sato, T., Hasegawa, S., Nakano, M., and Miyazaki, T., Soil water content in the vicinity of drying cracks having roots in clayey subsoil, *Soil Phys. Cond. Plant Growth 60*:24–27 (1990).

61. Kasubuchi, T., Heat conduction of soil, (in Japanese with English summary), *Bull. Natl. Inst. Agric. Sci. Ser. B33*:1–54 (1982).

62. Kasubuchi, T., Twin transient-state cylindrical-probe method for the determination of the thermal conductivity of soil, *Soil Sci. 124*:255–258 (1977).

63. Kasubuchi, T., Development of *in-situ* soil water measurement by heat-probe method, *Jpn. Agric. Res. Q. 26*:178–181 (1992).

9

Heterogeneity of Soils in Fields

I. HETEROGENEITY

Uniformity and heterogeneity of soil properties in a field are
defined in terms of the spatial variabilities of bulk density,
water content, matric potential, soil temperature, soil hardness, solute content, soil particle composition, etc. Since almost all of these soil properties are defined and measured
with respect to an element volume, the definition of uniformity and heterogeneity of soils depends on the element size or
measurement size.

Generally speaking, the larger the element sizes in a
field, the more the heterogeneity is averaged within each

element volume, resulting in more homogeneous properties over the field. On the other hand, the smaller the element sizes in the same field, the larger may be the differences between elements, resulting in more heterogeneous properties over the field.

When the scale of element is smaller than solid particles, as denoted by size 1 in Figure 9.1, where solid particles are denoted by hatched areas, the property of an element is determined by its location. In other words, this soil is never homogeneous when it is measured by an element smaller than solid particles. Even when the scale of element is a little larger than the smallest solid particles in a given soil, as denoted by size 2 in Figure 9.1, the soil is still heterogeneous because the fraction of each phase and the arrangement of solid particles in every element may be different from each other, resulting in different physical properties of each element. When the scale of element is so large that soil properties are averaged within each element, as denoted by size 3 in

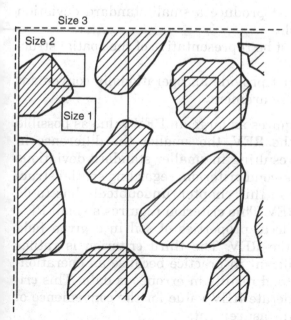

Figure 9.1 Dependence of soil heterogeneity on element sizes.

Figure 9.1, the soil properties of every element may be closed to each other, resulting in more homogeneous soil in the field.

The required sample size for defining the heterogeneity of soil in a field is generally termed the *representative elementary volume* (REV) or *representative elementary area* (REA). The difference between REV and REA is only the dimension involved, and hence no distinction between them is made in this book. Once the size of the REV (or REA) is defined in a given field, the average value of a measurement in each REV (or REA) is regarded as an independent sample of the field and the property of the field is determined by the set of these samples.

II. REPRESENTATIVE ELEMENTARY VOLUME

A. Criteria for REV

The size of the REV is generally determined by the following three criteria:

1. The REV must produce a small standard deviation among samples.
2. The REV must be representative of the spatial structure.
3. The REV must provide an operationally convenient method of measurement.

The first criterion requires as large an REV value as possible because the larger the REV, the smaller the difference in value among REVs, resulting in smaller standard deviations among samples. The second criterion, regarding spatial structure, explained later in this chapter, undoubtedly has to do with the size of the REV. This criterion requires a small REV value such that the local properties of soil in a given field are represented by the REV. The third criterion is rather technical but is significant in practice because an operationally cumbersome method results in erroneous data. This criterion requires a moderate REV value for the convenience of the person doing the measurement.

B. REV for Microstructural Heterogeneity

Cogels (1) investigated the representativity of a 1-cm^3 soil sample for evaluation of the heterogeneity of a well-structured Yolo loam field 40 by 40 m. He fixed a regular grid of 81 sampling locations for the first sampling and a complementary 13 locations for the second sampling in the test field as shown in Figure 9.2, where the locations of the first sampling are denoted by open circles and those of the second by solid circles. At each knot of the grid he extracted one undisturbed soil sample from within the top 5-cm layer.

The pore-size distributions of the two samples were measured using the mercury intrusion method. Cogels (1) classified the regularly extracted 81 samples into six sets, each made up of 12 to 20 samples, and calculated the average pore-size distribution curves independently for the six sets. The dashed lines in Figure 9.3 are the enveloping pore-size distribution curves of the six sets, and the solid curve is the approximate average distribution curve of the pore-size distribution. It is notable that the width of the two enveloping curves is very small. This means that even though the field was well structured, the microstructure of the soil was rather similar at each location.

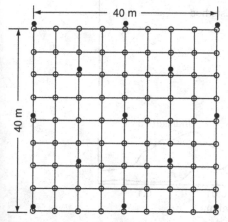

Figure 9.2 Regular grid of 81 sampling locations in a test field. (After Cogels, O. G., *Agric. Water Manage.* 6:203–211 (1983). With permission.)

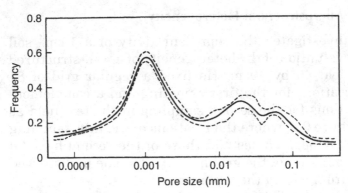

Figure 9.3 Pore-size distribution curves of six different extractions. (From Cogels, O. G., *Agric. Water Manage. 6*:203–211 (1983). With permission.)

The sensitivity of this measurement to difference in microstructures of the soil was examined by comparing the average pore-size distribution curve computed form all 94 undisturbed samples with that computed from 10 disturbed samples, as shown in Figure 9.4. The average pore-size distribution curve of the undisturbed samples has a large variance,

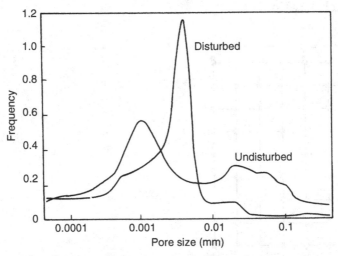

Figure 9.4 Average pore-size distribution curves of 94 undisturbed samples and 10 disturbed samples. (After Cogels, O. G., *Agric. Water Manage. 6*:203–211 (1983). With permission.)

whereas the curve of the disturbed samples has a sharp peak and a small variance. The fact that a reasonable difference appeared between these two curves supports the applicability of the mercury intrusion method for identification of soil microstructure.

Cogels (1) concluded from these results that the 1-cm^3 soil samples can be regarded as an REV for the estimation of soil microstructure in a field. Although such heterogeneities as macropores, cracks, and gravels in fields are not evaluated by this method and therefore the second criterion mentioned above is not satisfied in heterogeneous fields, this method is useful for evaluating the microstructures of relatively homogeneous soils in fields.

C. REV for Physical Properties of Soils in Fields

Tokunaga and Sato (2) and Sato and Tokunaga (3) investigated the use of REV for estimating bulk density and water content in an upland field. Instead of changing the sample size, they divided an area of 60 cm by 60 cm into 400 small test areas of equal size and extracted soil samples from all the small test areas. After measuring the bulk density and water content of each sample, they defined the variable test areas as designated by different-sized squares in Figure 9.5, which include different numbers of small test areas. Once the size of the variable test area was determined, each variable test area, within which bulk densities and water contents of soil samples were averaged, was designed so as to cover all the test fields by shifting the locations one by one.

After averaging the bulk densities and water contents of the variable test area, the coefficient of variation CV (%), defined by

$$CV = \frac{SD}{\langle m \rangle} \times 100 \tag{9.1}$$

was calculated, where $\langle m \rangle$ is the average value and SD is the standard deviation of the values of the variable test area. Figure 9.6 shows the values of CV of bulk densities and water contents as functions of the sizes of the variable test

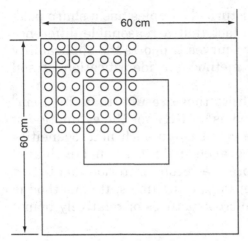

Figure 9.5 Design of variable test area that contains different numbers of samples arranged regularly. (From Sato, T. and Tokunaga, K., *Trans. Jpn. Soc. Irrig. Drain. Reclam. Eng.* *63*:1–7 (1976). With permission.)

area. Clearly, the coefficient of variation decreases with increase in the variable test area. Depending on Figure 9.6, the variable test area in a field is required to cover at least 1600 cm^2 to obtain a coefficient of variation of less than 1%. However, taking account of the operational convenience, as

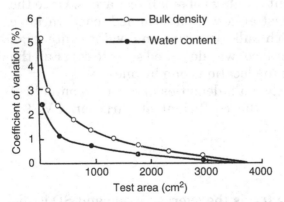

Figure 9.6 Coefficient of variation of bulk densities and water contents as functions of the test area. (After Sato, T. and Tokunaga, K., *Trans. Jpn. Soc. Irrig. Drain. Reclam. Eng.* *63*:1–7 (1976). With permission.)

given by criterion 3, Sato and Tokunaga (3) recommended use of a cylindrical sampler that covers an area of 20 cm² and whose volume is 100 cm³.

Rice and Bowman (4) investigated the difference between a small sampler and a large sampler, as illustrated in Figure 9.7 for the estimation of solute transport in a field of sandy loam. The smaller cylindrical sampler had a diameter of 2.2 cm; that of the larger sampler was 10.3 cm. They concluded experimentally that the sample-size effects were far less than the spatial structural effects in estimating Br⁻ concentration in the test field and, taking account of the operational convenience given by criterion 3, that the diameter of 2.2 cm was satisfactory for use with REV.

D. REV for Hydraulic Permeability of Soil with Macropores

Lauren et al. (5) investigated the REV for the prediction of saturated hydraulic conductivities of silty clay loam in a field by suing *in situ* columns. The soil had a structure that is moderate to strong, subangular blocky, with medium prisms,

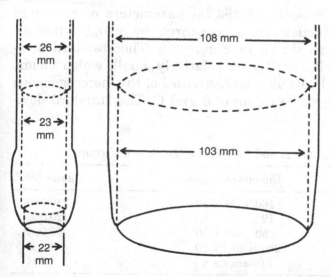

Figure 9.7 Small and large soil samplers used by Rice, R. C. and Bowman, R. S., *Soil Sci.* *146*(2):108–112 (1988). With permission.

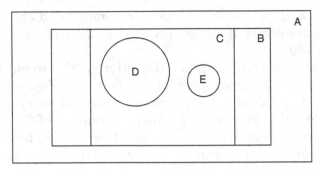

Figure 9.8 Five types of *in situ* field columns used by Lauren, J. G., Wagnent, R. J., Bouma, J., and Wosten, J. H. M., *Soil Sci.* 145(1):20–28 (1988). With permission.

firm, and very plastic. They used five types of *in situ* columns of different sizes, as shown in Figure 9.8. The dimensions of these columns are given in Table 9.1, where A is the largest sample and E is the smallest sample. The saturated hydraulic conductivities were measured by ponding water in each column without disturbing the field soil at 37 equally spaced measurement locations. The size of the REV was determined as follows.

Table 9.2 lists the statistical parameters of saturated hydraulic conductivity K_{sat} measured by using different-sized columns as shown in Figure 9.8. The standard deviations of sizes B and C were relatively small, which means that saturated hydraulic conductivities of this heterogeneous soil were averaged in columns B and C. The standard devi-

Table 9.1 Dimensions and Volumes of *In Situ* Columns

Column	Dimensions (cm)	Volume (cm^3)
A	160 × 75 × 20	240,000
B	120 × 75 × 20	120,000
C	50 × 50 × 20	50,000
D	20 (diam.) × 20	6,283
E	7 (diam.) × 6	884

Source: Lauren, J. G., Wagnent, R. J., Bouma, J., and Wosten, J. H. M., *Soil Sci.* 145(1):20–28 (1988). With permission.

Table 9.2 Statistical Parameters for K_{sat} Measured by Different-Sized *In Situ* Columns

Size	cm day^{-1} Mean	Mode	Median	Standard deviation	Coefficient of variation (%)	Number of samples
A	21.3	10.3	16.6	16.9	79	37
B	13.7	6.4	10.7	11.0	81	36
C	14.4	6.3	10.9	12.5	96	37
D	36.6	6.3	20.3	54.9	150	37
E	34.5	4.8	16.3	64.0	186	35

Source: Lauren, J. G., Wagnent, R. J., Bouma, J., and Wosten, J. H. M., *Soil Sci.* *145*(1):20–28 (1988). With permission.

ation of the largest column (A) was a little higher, due to the operational problems in measurements. The standard deviations of sizes D and E were so large, probably due to the large size heterogeneity of soils in the columns, that these REV values were not acceptable. By using the additional criteria 2 and 3 mentioned above, they concluded that column size C, whose size is 50 cm by 50 cm in width and 20 cm in depth, was the optimal REV for measuring saturated hydraulic conductivity *in situ*.

E. Other REVs

There are many other examples presenting the adaptable REV for each purpose specified, but the general theory for giving the REV in fields is not established yet. The tentatively acceptable empirical REV values are given in Table 9.3, where the sizes are designated by representative REV length (m).

Since water balance in fields with cracks is dominated mainly by the hydraulic properties of crack networks rather than by the hydraulic properties of the soil matrix, the size of the REV depends on the degree of cracks and their density. According to the field study conducted by Inoue et al. (6), the hydraulic property of the field with highly developed cracks was strongly affected by the sizes of test blocks when they were smaller than 5 m in length. Adopting the three criteria

Table 9.3   Empirical REV Designated by Representative Length (m)

	REV size	Subject	References
Large	REV > 1000 (m)	Hydrological modeling of a river basin	7
	REV > 5 (m)	Water balance in field with cracks	6
	REV > 0.5 (m)	Saturated hydraulic conductivity of soil with macropores	5
	REV > 0.05 (m)	Bulk density, water content, and solute concentration	2–4
Small	REV > 0.01 (m)	Microstructures	1

given above, REVs larger than 5 m may be operationally cumbersome. The empirical size of an REV for the evaluation of water balance with cracks is therefore estimated to be 5 m in Table 9.3.

Larger sizes are required as the REA for hydrologic modeling in river catchments, because areal fluctuations of rainfall, runoff, infiltration, and evaporation must be averaged within each REA in order to estimate water balance and river runoff as accurately as possible. Wood et al. (7) investigated the size of the REA in the Coweeta River experimental basin (an area of 17 km^2) by dividing the basin into several subcatchments. In their preliminary investigation, they suggested that 1 km^2 is a promising REA for the basin.

III. SIMILAR-MEDIA CONCEPT AND SCALING

A. Characteristic Length and Scaling

One of the smallest REVs is the one proposed by Miller and Miller (8), who illustrated two similar media in a state similar to that shown in Figure 9.9(a) and (b), where the two characteristic lengths, λ_a and λ_b, connect corresponding points in the two similar media. On the other hand, a medium composed of the same solid particles as in medium (a) but different in arrangement, as illustrated in Figure 9.9(c), is no longer

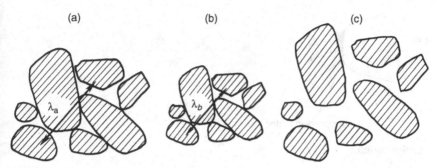

(a) (b) (c)

Figure 9.9 Similar and nonsimilar media.

similar to media (a) and (b). In such a sense, as noted by Miller and Miller, the probability of the occurrence of detailed similarity throughout the microscopic shapes of two media is practically zero.

The similar-media concept has provided a significant opportunity to determine the physical properties of different soils theoretically and systematically, while the discrepancy between the similarity of conceptual porous media and the nonsimilarity of actual soils has limited the applicability of this concept.

The term *scaling* or *scaling theory* is used when the character of one porous medium is related to that of a similar porous medium by means of the parameters λ_a and λ_b or by their ratio, λ_a/λ_b. The relation of retentivities of water, hydraulic conductivities, and other physical properties between similar porous bodies can be deduced through the scaling procedures as described below.

B. Similarity of Pressure and Matric Head of Water

When volumetric water contents are equal in two similar porous media, the shapes of water held in them will also be similar. Figure 9.10(a) and (b) illustrates the relation between characteristic lengths, denoted by λ_a and λ_b, and the shapes of suspended water held at the contacts of the two soil particles. The shapes and pressures of the suspended water are characterized by using the radii of surface curvatures of a

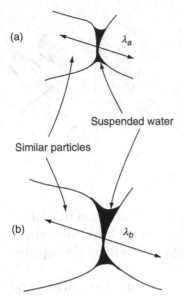

(a)

λ_a

Suspended water

Similar particles

(b)

λ_b

Figure 9.10   Similar water suspended at contacts of similar media.

liquid meniscus r_{1a} and r_{2a} in medium (a) and r_{1b} and r_{2b} in medium (b).

It is clear that these radii of surface curvatures are related geometrically to each other by

$$r_{1a} = \alpha r_{1b}$$
$$r_{2a} = \alpha r_{2b}$$

(9.2)

where α is the geometrical similarity ratio, defined by

$$\alpha = \frac{\lambda_a}{\lambda_b}$$

(9.3)

This relation is used to express the pressure and matric head of water retained in similar media.

When suspended water held as illustrated by Figure 9.10(a) is in equilibrium, the difference between the pressure in the liquid phase p (N m^{-2}) and the pressure in the gas phase p_0 (N m^{-2}) is given by the classical equation

$$\Delta p_a = \sigma_w \left(\frac{1}{r_{1a}} + \frac{1}{r_{2a}} \right) \cos \delta$$

(9.4)

where $\Delta p_a = p_0 - p$, δ is the contact angle of water with the solid surface, and σ_w is the surface tension of water (N m^{-1}). The algebraic average of the curvatures of the two curves of a liquid meniscus is given by

$$\frac{1}{\langle r_a \rangle} = \frac{1}{2} \left(\frac{1}{r_{1a}} + \frac{1}{r_{2a}} \right) \tag{9.5}$$

where $1/\langle r_a \rangle$ is the average curvature. Using Equation (9.5), Equation (9.4) is simplified into

$$\Delta p_a = \cos \delta \frac{2\sigma_w}{\langle r_a \rangle} \tag{9.6}$$

In the same manner, the averaged curvature of meniscus illustrated by Figure 9.10(b) is given by

$$\frac{1}{\langle r_b \rangle} = \frac{1}{2} \left(\frac{1}{r_{1b}} + \frac{1}{r_{2b}} \right) \tag{9.7}$$

Taking account of the relation given by Equation (9.2), $\langle r_a \rangle$ is easily related to $\langle r_b \rangle$ as

$$\frac{1}{\langle r_a \rangle} = \frac{1}{\alpha \langle r_b \rangle} \tag{9.8}$$

Hence,

$$\Delta p_a = \frac{1}{\alpha} \Delta p_b \tag{9.9}$$

is obtained where Δp_b is the pressure differences in both sides of meniscus in medium (b). Equation (9.9) is the similarity of pressure of water suspended in a similar medium.

Assuming that the matric head of water in these media is determined solely by the capillary pressure, we also obtain

$$\psi_a = \frac{1}{\alpha} \psi_b \tag{9.10}$$

or, alternatively,

$$\lambda_b \psi_b = \lambda_a \psi_a \tag{9.11}$$

where ψ_a and ψ_b are the matric head (m) of water in media (a) and (b), respectively, in Figure 9.10.

C.   Similarity of Hydraulic Conductivity

As for saturated hydraulic conductivity K_s (m s^{-1}), Miller and Miller (8) stated that K_s must vary directly with λ^2, due to a mathematical requirement in the linearized Navier–Stokes equation at the microscopic scale. Poiseuille's law provides another way and the same result in finding the similarity of K_s.

The quantity of saturated laminar flow in a narrow pipe (Figure 9.11) is given by Poiseuille's law,

$$Q = \frac{\pi r^4}{8\mu} \frac{\Delta p}{\Delta x} \tag{9.12}$$

where Q is the flow rate (m^3 s^{-1}), r is the radius of the pipe (m), μ is the coefficient of viscosity (Pa s), and $\Delta p/\Delta x$ is the pressure gradient along the x-axis (Pa m^{-1}). When Darcy's law is applicable to the flow in this pipe, the flux of water q (m^3 m^{-2} s^{-1}) is given by

$$q = -\frac{K_s}{\rho_w g} \frac{\Delta p}{\Delta x} \tag{9.13}$$

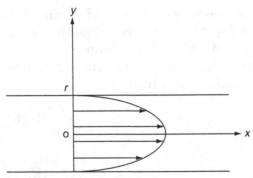

Figure 9.11   Velocity distribution of saturated laminar flow in a narrow pipe.

where K_s is the saturated hydraulic conductivity of the pipe (m s^{-1}), ρ_w is the density of water (kg m^{-3}), and g is the gravity constant (m s^{-2}). Taking account of the relation that $q = Q/(\pi r^2)$, K_s is defined as

$$K_s = \frac{\rho_w g r^2}{8\mu} \tag{9.14}$$

which indicates that K_s is proportional to r^2. Extending this definition of K_s into similar bundles of pipes saturated with water as given in Figure 9.12, we obtain the similarity of K_s between bundles (a) and (b) as

$$K_{sa} = \alpha^2 K_{sb} \tag{9.15}$$

or, alternatively,

$$\frac{K_{sb}}{\lambda_b^2} = \frac{K_{sa}}{\lambda_a^2} \tag{9.16}$$

where K_{sa} and K_{sb} are the saturated hydraulic conductivities of each bundle and α is their similarity ratio.

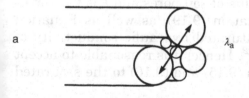

a

λ_a

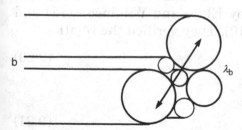

b

λ_b

Figure 9.12 Extension of similar-media concept to bundles of pipes saturated with water.

The extension of the similarity relations (9.15) and (9.16) of bundles of pipes to saturated similar porous media is conducted with the help of models of saturated hydraulic conductivities of soils. One of the best-known models, proposed by Childs and Collis-George (9), is given by

$$K_{\mathrm{s}} = \frac{\rho_{\mathrm{w}}g}{8\mu} \int_0^r \int_0^r r^2 F(r)\, dr\, F(r)\, dr \qquad (9.17)$$

in which r is the radius of soil pores. $F(r)$ is the pore-size distribution function, defined by

$$n = \int_0^{r_{\max}} F(r)\, dr \qquad (9.18)$$

where n is the porosity of the soil and $r_{\max}$ is the maximum radius of the pore in the soil. Equation (9.17) implies that the saturated hydraulic conductivity of soil is proportional to r^2.

Another model, presented by Campbell (10) after Scheidegger (11), is given by

$$K_{\mathrm{s}} = \frac{\rho_{\mathrm{w}}gnr^2}{\mu\xi} \qquad (9.19)$$

where r is the average radius of soil pores and ξ is the tortuosity (dimensionless). Equation (9.19), as well as Equation (9.17), implies that the saturated hydraulic conductivity of the soil is proportional to r^2. Hence, it is reasonable to accept the extension of Equations (9.15) and (9.16) to the saturated similar soils.

Further extension of similarity relations (9.15) and (9.16) to unsaturated similar soil as shown in Figure 9.10(a) and (b) was given experimentally by Klute and Wilkinson (12) and theoretically by Campbell (10). They verified the relation

$$K_{\mathrm{a}}(\psi_{\mathrm{a}}) = \alpha^2 K_{\mathrm{b}}(\psi_{\mathrm{b}}) \qquad (9.20)$$

or, alternatively,

$$\frac{K_{\mathrm{b}}(\psi_{\mathrm{b}})}{\lambda_{\mathrm{b}}^2} = \frac{K_{\mathrm{a}}(\psi_{\mathrm{a}})}{\lambda_{\mathrm{a}}^2} \qquad (9.21)$$

for unsaturated similar soils, where $K_a(\psi_a)$ and $K_b(\psi_b)$ are unsaturated conductivities of media (a) and (b) in Figure 9.10 as functions of matric heads ψ_a and ψ_b, respectively. Note that ψ_a and ψ_b are the matric heads of the same value of volumetric water content in similar porous bodies, and therefore their values are different. Inevitably, the values of $K_a(\psi)$ and $K_b(\psi)$ at a given matric head ψ are not equal.

D. Verification and Limitation of Similar-Media Concept

Klute and Wilkinson (12) and Wilkinson and Klute (13) verified experimentally that scaling theory was applicable to the five sand fractions, which had similar particle-size frequency curves but had different average particle sizes. They used the arithmetic means of the upper and lower sieve sizes as the characteristic microscopic lengths λ. By substituting individual values of λ into Equations (9.20) and (9.21), they reduced the moisture characteristic curves and unsaturated hydraulic conductivities of these five sand fractions. Figure 9.13 illustrates the conceptual particle-size distribution curves of fractions a and b having similar curves but different average sizes. By using the procedure of Klute and Wilkinson (12) mentioned above, the characteristic lengths are given by

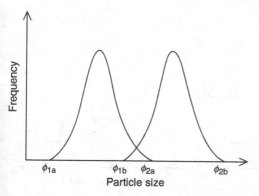

Figure 9.13 Hypothetical particle-size distribution curves of fractions a and b having similar curves but different average sizes.

$$\lambda_a = \frac{\phi_{1a} + \phi_{2a}}{2} \qquad\qquad (9.22)$$

$$\lambda_b = \frac{\phi_{1b} + \phi_{2b}}{2} \qquad\qquad (9.23)$$

where λ_a and λ_b are the characteristic lengths of fractions a and b, respectively, ϕ_{1a} and ϕ_{2a} are the minimum and maximum particle sizes of fraction "a" and ϕ_{1b} and ϕ_{2b} are those of fraction "b." Assuming that fractions a and b are similar media, scaling theory predicts that the original soil moisture characteristic curves (left-hand side Figure 9.14) are reduced to one curve (right-hand side of Figure 9.14), and the original unsaturated hydraulic conductivities (left-hand side of Figure 9.15) are reduced to one curve (right-hand side of Figure 9.15), respectively. When the hydraulic conductivity of additional similar fraction "c" is obtained, the function will be reduced to one curve by the scaling procedure.

In general, the agreements of the reduced moisture characteristic curves and reduced unsaturated hydraulic conductivities were good in the experiment by Klute and Wilkinson (12). However, disagreement of reduced moisture characteristic curves between experiment and theory was apparent in their data, particularly when the volumetric water content was greater than 0.30 cm^3 cm^{-3}.

Elrick et al. (14) also investigated experimentally the applicability of the scaling theory to soil by using different

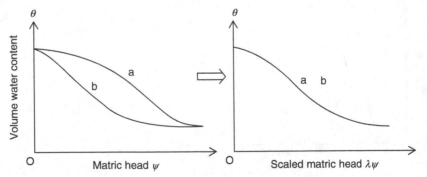

Figure 9.14 Original soil moisture characteristic curves and the reduced soil moisture characteristic curve by scaling.

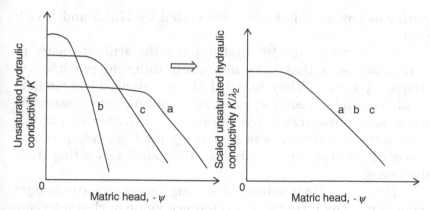

Figure 9.15 Original hydraulic conductivity curves and the reduced hydraulic conductivity curve by scaling.

liquids (water and butyl alcohol) in the same medium and by using one liquid (water) in different (but similar) media. An alundum mixture (probably classified as loam), clay-free silt loam, and clay-included silt loam were chosen as the similar media. After analyzing the data of hysteresis loops for liquid characteristic curves and unsaturated hydraulic conductivity curves of the same medium, and analyzing wetting front advancement during infiltration and the change in drainage rate with time in the different media, they concluded that scaling theory worked well when the medium was clean sand, but much less well when the amount of colloid increased in the media.

Taking account of all the experiments exemplified above and the many other experimental works, it is likely that as remarked appropriately by Tillotson and Nielsen (15), application of scaling theory is restricted to use in sand or sandy soils.

E. Scaling of Natural Soils

Many efforts have been made in the 1970s to apply the similar-media concept further to a variety of natural soils (e.g., Reichardt et al. (16,17), Peck et al. (18) Warrick et al. (19), Warrick and Amoozegar-Fard (20), Sharma and Luxmoore (21), Simmons et al. (22), Russo and Bresler (23), etc.). Recent contributions on this topic were edited by Hillel and Elrick (24).

Reichardt et al. (16) first applied the scaling theory to nonsimilar soils that have absolutely different particle-size frequency curves. They found that even though the similar-media concept is poorly applicable to soil moisture characteristics and unsaturated hydraulic conductivities of natural field soils, the theory was promising for the scaling of soil water diffusivities and infiltration data such as wetting-front distances.

Peck et al. (18) applied the scaling theory to water budget modeling. They introduced a reference value of characteristic length λ^* and represented the physical properties of their watershed by the variable characteristic length λ, which was the function of location in the watershed.

Many other investigations on the applicability of the similar-media concept and associating scaling theory to natural soils have been directed to adopting the characteristic length λ as the REV (REA or representative element length [REL] may be preferable in some cases) for their particular purposes. Unfortunately, the restriction of scaling theory and difficulties in finding appropriate values of λ are providing limiting success up to now.

IV. NONSIMILAR MEDIA CONCEPT AND SCALING

A. Characteristic Lengths and Shape Factor

Since the assumption of similarity for all natural soils is, as exemplified in Figure 9.9, too restrictive to proceed with the scaling technique, Miyazaki (25) proposed a nonsimilar concept (NSMC) model.

The NSMC model consists of a characteristic length S for the solid phase and a characteristic length d for the pore space, with which an element of a given soil is given by $(S + d)^3$. These characteristic lengths are, however, not directly measurable but are defined as representative lengths for both phases. By using S and d, the volume V and the mass M of a given soil are defined by

$$V = N(S + d)^3 \tag{9.24}$$

$$M = N\tau\rho_s S^3 \tag{9.25}$$

where ρ_s is the soil particle density and N is the number of elements included in the given soil. The parameter τ is the shape factor of the solid phase and is defined as the ratio of the substantial volume of the solid phase to the volume S^3.

The shape factor τ is unity when the shapes of soil particles are all cubes as shown in Figure 9.16(a), while it is equal to $\pi/6$ when their shapes are all spheres as shown in Figure

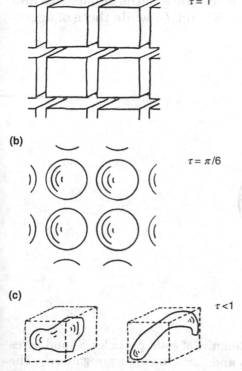

(a) $\tau = 1$

(b) $\tau = \pi/6$

(c) $\tau < 1$

Figure 9.16 The shape factor of hypothetical porous media made of (a) cubes, (b) spheres, and (c) irregular particles.

9.16(b). Since natural soil particles are all different as shown in Figure 9.16(c), the substantial volume of the solid phase in an element is generally given by τS^3 and τ is equal to or less than one.

B. Deduced Bulk Density

Once the values of S, d, and τ are determined for a soil, its bulk density ρ_b is defined by

$$\rho_b = \frac{M}{V} = \tau \rho_s \left(\frac{S}{S+d}\right)^3 \tag{9.26}$$

C. NSMC Model for Aggregated Soils

Figure 9.17 shows the application of the NSMC model for an aggregated soil. The characteristic lengths of particles and pores within an aggregate are S_1 and d_1, while those of aggre-

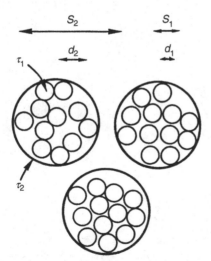

Figure 9.17 Characteristic lengths of solid particles S_1 and pores d_1, and those of aggregates S_2 and pores among aggregates d_2. The shape factors τ_1 and τ_2 are associated with S_1 and S_2. (After Miyazaki, T., *Soil Sci. 161*:484–490 (1996). With permission.)

gates and pores among aggregates are S_2 and d_2. The shape factors for soil particles and aggregates are τ_1 and τ_2, respectively.

Based on the general definition given above, the bulk density of one aggregate ρ_{ag} is defined as

$$\rho_{ag} = \tau_1 \rho_s \left(\frac{S_1}{S_1 + d_1}\right)^3 \tag{9.27}$$

and the bulk density of this soil is

$$\rho_b = \tau_2 \rho_{ag} \left(\frac{S}{S_2 + d_2}\right)^3 \tag{9.28}$$

Substituting (9.27) into (9.28), we obtain

$$\rho_b = \tau_1 \tau_2 \rho_s \left(\frac{S_1}{S_1 + d_1}\right)^3 \left(\frac{S_2}{S_2 + d_2}\right)^3 \tag{9.29}$$

If we define the "apparent shape factor of aggregate" τ^* by

$$\tau^* = \left(\frac{S_1}{S_1 + d_1}\right)^3 \tau_1 \tau_2 \tag{9.30}$$

then the bulk density is

$$\rho_b = \tau^* \rho_s \left(\frac{S_2}{S_2 + d_2}\right)^3 \tag{9.31}$$

Equation (9.31) is apparently the same as Equation (9.26) except for the definition of the shape factors.

Eventually, the bulk density of any soil is defined in the form of Equation (9.26) in which the shape factor τ is the ratio of the solid phase to the volume V^3 for dispersed soils and is the apparent shape factor for aggregated soils. This equation is applicable to soils that have the third or more aggregations. However, it is noted again that the characteristic lengths S, d, S_1, d_1, S_2, and d_2 are not measurable values.

D. Bulk Density Dependence of Air Entry Suction and Hydraulic Conductivity

1. NSMC Model

Even though the characteristic lengths in the NSMC model are not measurable, it is reasonable to assume that air entry suction h_e is inversely proportional to d, such that

$$\frac{h'_e}{h_e} = \frac{d}{d'} \tag{9.32}$$

where h'_e is the air entry suction of the same soil whose bulk density is ρ'_b. By using Equation (9.26) for the soil with bulk densities ρ_b and ρ'_b, respectively, the relation

$$\frac{h'_e}{h_e} = \frac{(\tau\rho_s/\rho_b)^{1/3}-1}{(\tau\rho_s/\rho'_e)^{1/3}-1} \tag{9.33}$$

According to the discussion given in the previous section (see Equations (9.17) and (9.19)), the saturated hydraulic conductivity of a soil is proportional to the second power of the average radius r of soil pores and, therefore, may be proportional to the second power of the characteristic length d as

$$\frac{K'_s}{K_s} = \left(\frac{r'}{r}\right)^2 = \left(\frac{d'}{d}\right)^2 \tag{9.34}$$

This leads us to the equation

$$\frac{K'_s}{K_s} = \left[\frac{(\tau\rho_s/\rho'_b)^{1/3}-1}{(\tau\rho_s/\rho_b)^{1/3}-1}\right]^2 \tag{9.35}$$

2. Campbell's Method

Campbell and Shiozawa (26,27) proposed an empirical equation for estimating the bulk density dependence of air entry suction h_e and saturated hydraulic conductivity K_s as

$$h_e = h_{es}\left(\frac{\rho_b}{1.3}\right)^{0.67b} \tag{9.36}$$

$$K_s = 4 \times 10^{-3} \left(\frac{1.3}{\rho_b}\right)^{1.3b} \exp(-6.9m_c - 3.7m_s) \tag{9.37}$$

where h_{es} is an air entry suction for the standard bulk density of 1.3 Mg m^{-3}, b is the power of the soil moisture characteristic function, m_c is the clay mass fraction, and m_s is the silt mass fraction. The parameter b is determined by

$$b = -2\psi_{es} + 0.2\sigma_g \tag{9.38}$$

where ϕ_{es} (J kg^{-1}) is the air entry potential for standard bulk density of 1.3 Mg m^{-3} and σ_g (mm) is the geometric standard deviation of particle diameter. The value of ϕ_{es} is related to the geometric mean particle diameter d_g (mm) by

$$\psi_{es} = -0.5 d_g^{-1/2} \tag{9.39}$$

In short, Equations (9.36) and (9.37) estimate the air entry suction and the hydraulic conductivity from the particle-size distribution data and the moisture characteristic curve of the soil.

It is noted that, since h_{es}, b, m_c, and m_s are all dependent on the particle-size distribution and independent of the soil bulk density, the equations

$$\frac{h_e'}{h_e} = \left(\frac{\rho_b}{\rho_b'}\right)^{0.67b} \tag{9.40}$$

and

$$\frac{K_s'}{K_s} = \left(\frac{\rho_b}{\rho_b'}\right)^{1.3b} \tag{9.41}$$

are deduced, where h_e' and K_s' are again associated with the bulk density ρ_b'. The procedure to find the b-value is given in the appendix.

3. Kozeny–Carman Equation

The well-known Kozeny–Carman equation

$$K_s = \frac{n^3}{ca^2(1-n)^2} \tag{9.42}$$

gives another bulk density scaling of hydraulic conductivity, where n is the porosity, c is the so-called "Kozeny constant," and a is the specific surface (28). According to Carman, $c = 5$ is the best value for many soils.

By applying Equation (9.42) to the different bulk densities ρ_b and ρ'_b of the same soil and by using the relation between porosity and bulk density, $n = 1 - \rho_b/\rho_s$, the bulk density scaling procedure is performed as

$$\frac{K'_s}{K_s} = \left(\frac{\rho_b}{\rho'_b}\right)^2 \left(\frac{\rho_s - \rho'_b}{\rho_s - \rho_b}\right)^3 \tag{9.43}$$

4. Comparison of Models

Table 9.4 shows the five soils of different textures, namely sand, sandy loam, light clay, loamy sand, and clay loam, to be compared. The b-value used in Campbell's method is included in the table.

Figure 9.18 shows the measured and predicted air entry suctions as a function of bulk density of a volcanic ash origin sandy loam. Figure 9.19 shows the measured and predicted hydraulic conductivities of the same soil. The value of $\tau^* = 0.21$ is suitable for both predictions. The extremely low bulk densities for this soil are attributed to the high porosities, occasionally more than 80%.

Table 9.4 Mass Fractions of Texture Classes, Soil Particle Densities, and b Values

Soil type	Fractions of texture class			Particle density ρ_s (g cm^{-3})	b Values in Equations (9.40) and (9.41)
	Clay	Silt	Sand		
Volcanic ash origin sandy loam	0.11	0.16	0.73	2.63	4.08
Toyoura sand	0	0	1	2.63	1.19
Kunigami maji light clay	0.34	0.34	0.32	2.74	9.39
Norfolk loamy sand	0.048	0.11	0.84	—	2.64
Fukaya clay loam	0.22	0.31	0.48	2.71	12.2

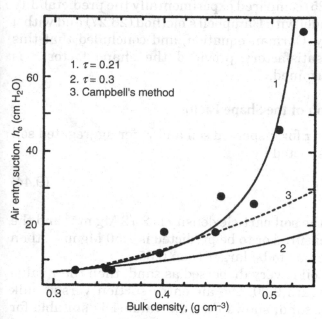

Figure 9.18 Predicted and measured air entry suctions of volcanic ash origin sandy loam. (After Miyazaki, T., *Soil Sci. 161*:484–490 (1996). With permission.)

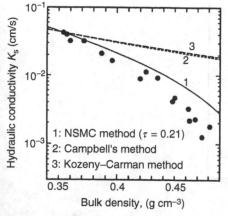

Figure 9.19 Predicted and measured hydraulic conductivities of volcanic ash origin sandy loam. (After Miyazaki, T., *Soil Sci. 161*:484–490 (1996). With permission.)

Miyazaki (25) compared experimentally the predictability of the NSMC model with Campbell's method (26,27) and with a modified Kozeny–Carman equation, and concluded that this new model is satisfactory provided the shape factor τ is adequately determined.

E. Determination of the Shape Factor

The shape factor τ for dispersed soil and τ^* for aggregated soil are basically restricted by

$$\frac{\rho_b}{\rho_s} < \tau, \quad \tau^* \leq 1 \tag{9.44}$$

For example, if the soil particle density is $2.72\ \text{Mg m}^{-3}$ and the bulk density measured or to be predicted is $1.50\ \text{Mg m}^{-3}$, then the τ or τ^* value has to be larger than 0.55.

When the soil is very dispersed as sand, then the τ value is close to 1. Figure 9.20, the air entry suction versus bulk density Toyoura sand, shows that the value 1 is suitable for the bulk density scaling of h_e. Figure 9.21, the saturated hydraulic conductivity versus bulk density of this sand,

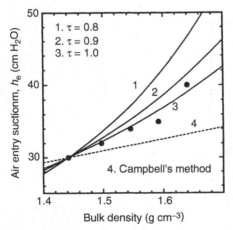

Figure 9.20   Predicted and measured air entry suctions of Toyoura sand. (After Miyazaki, T., *Soil Sci. 161*:484–490 (1996). With permission.)

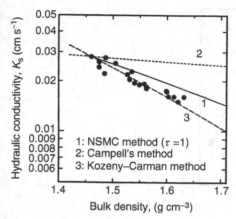

Figure 9.21 Predicted and measured hydraulic conductivities of Toyoura sand. (After Miyazaki, T., *Soil Sci. 161*:484–490 (1996). With permission.)

shows the value of τ is close to 1 and 0.9 may be more suitable for the bulk density scaling of K_s.

When the soil is aggregated, the τ^* value is close to ρ_b/ρ_s. Figure 9.22, h_e versus ρ_b of Kunigami maji light clay, shows that 0.5 is suitable for the bulk density scaling and this value is close to the least value of measured $\rho_b/\rho_s = 1.1/2.74 = 0.40$.

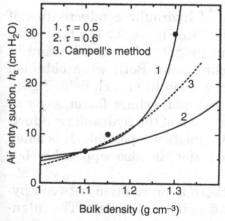

Figure 9.22 Predicted and measured air entry suctions of Kunigami maji light clay. (After Miyazaki, T., *Soil Sci. 161*:484–490 (1996). With permission.)

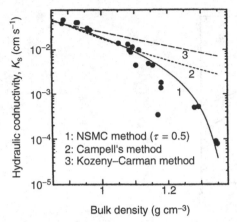

Bulk density (g cm⁻³)

Figure 9.23 Predicted and measured hydraulic conductivities of Kinigami maji light clay. (After Miyazaki, T., *Soil Sci. 161*:484–490 (1996). With permission.)

Figure 9.23, K_s versus ρ_b of this soil, shows that $\tau^* = 0.5$ is quite suitable for the prediction.

The volcanic ash soil is also an aggregated soil whose least value of measured ρ_b/ρ_s is $0.32/2.63 = 0.12$. The value $\tau^* = 0.21$ determined in Figure 9.18 and Figure 9.19 is a little larger than this lower limit but far from 1.

F. Application of the NSMC Model in Fields

The bulk density dependence of hydraulic conductivities of upland field soils is exemplified in Figure 9.24, whose data are from Cassel (29), for nonaggregated loamy sand and in Figure 9.25 for aggregated clay loam. Both were obtained from various locations and various depths of each field. Figure 9.24 shows that if we chose an adequate shape factor, say $\tau = 0.75$ to 0.80, then a rough estimation of the hydraulic conductivity as a function of their bulk densities is possible. It is noted that the Kozeny–Carman equation is also applicable but Campbell's method is not.

Figure 9.25 shows that there is no relation between hydraulic conductivity and bulk density in this field. The intensive plowing may have destroyed even the partial similarity of soil pores in the same field. The NSMC model and any other model are not applicable in such a field.

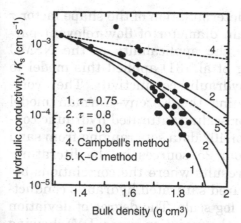

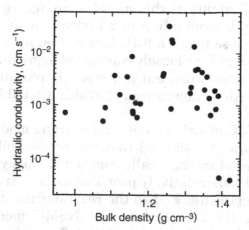

Figure 9.24 Predicted and measured hydraulic conductivities of Norfolk loamy sand. (After Miyazaki, T., *Soil Sci. 161*:484–490 (1996). With permission.)

Figure 9.25 Hydraulic conductivity versus soil bulk density of clay loam sampled from Fukaya cultivated upland field.

G. Applicability and Limitation of the NSMC Model

The applicability of the NSMC model to sandy soil, whose shape factor is almost always 1, is sufficiently high. Zhuang et al. (30) approved the applicability of the NSMC model to other soils whose shape factors are less than 1.

They emphasized that the determination of the shape factor τ by using the geometric mean diameter of flow-relevant particles could fairly improve the applicability of the NSMC model. In addition, Zhuang et al. (31) applied this model to the scaling of saturated hydraulic conductivity. They compared eight models, including the Kozeny–Carman model and the Campbell model, for scaling saturated hydraulic conductivity using a database of 402 data sets ranging from sand to heavy clay collected from 25 sources in the literature. Figure 9.26 shows all the results where the correlations between estimated and measured saturated hydraulic conductivities are represented in a log scale. The degree of deviation of the correlation is given by the deviation times (DV) showing the smallest value of 3.17 for the NSMC model. They thus concluded that the NSMC model performed the best scaling out of the eight models (31).

The applicability of this model to an aggregated soil depends on the determination of an adequate shape factor τ^*. The τ^* value close to but a little larger than ρ_b/ρ_s, where ρ_b is the bulk density of the loosely compacted aggregated soil, may give a rough estimation, but it is strongly recommended that the $h_e-\rho_b$ relation is measured by which a reliable τ^* value is determined.

The NSMC model is a new and growing model. Miyazaki and Nishimura (32) showed that the predictability of K_s as functions of ρ_b of mechanically compacted clayey soils by the NSMC model were relatively poor, due to the extreme destruction of the aggregates and to the rearrangement of clay particles during the compaction. The NSMC model for such compacted soils may be improved by defining the shape factor as a function of the bulk density or compaction energy.

H. Extension of the NSMC Model to Unsaturated Soils

Zhuang et al. (33,34) further extended and modified the NSMC model to use for the unsaturated soils. They found good applicability of this model both in estimating the unknown soil water retention characteristics for different tex-

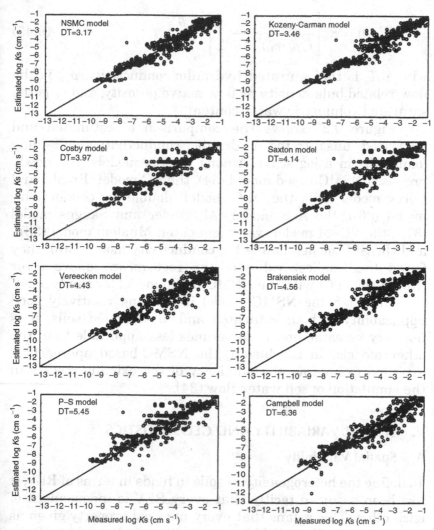

Figure 9.26 Comparison of values of log K_s measured and estimated by means of the eight models. (After Zhuang, J., Nakayama, K., Yu, G. R., and Miyazaki, T., *Soil Sci. 165*:718–727 (2000). With permission.)

tural soils and in estimating the unsaturated hydraulic conductivities of different dry bulk densities. The newly proposed NSMC-based equation for unsaturated hydraulic conductivity, K_{us}, is given by (34)

$$K_{us}(\theta) = K_s \left[\frac{(\tau\rho_s/\rho_{ub})^{1/3} - 1}{(\tau\rho_s/\rho_b)^{1/3} - 1} \right]^2 \left(\frac{\theta}{\theta_s} \right)^{2b} \tag{9.45}$$

where K_s is the saturated hydraulic conductivity, ρ_{ub} is the flow-related bulk density or flow-active porosity, and θ_s is the saturated volumetric water content.

Figure 9.27 shows the comparison of estimated and measured unsaturated hydraulic conductivities of sandy loam soils in a log scale. Among the six models, the newly proposed NSMC-based model (34), the BC model (Brooks and Corey model (35)), the DLC model (double log conductivity model (36)), the A–S model (Alexander and Skaggs model (37)), the VG–M model (van Genuchten-Mualem model (38)), and the SLC model (single log conductivity model (36)), they found the smallest value of the root mean square error (RMSE) of 1.001 with the NSMC-based model for sandy loam.

Although the NSMC model is showing relatively good applicability to both saturated and unsaturated soils, it is not very versatile and is sometimes less applicable than the other models. In conclusion, the NSMC-based approach is expected to improve the efficiency of the existing models in the simulation of soil water flow (34).

V.   SPATIAL VARIABILITY AND GEOSTATISTICS

A.   Spatial Variability

To define the heterogeneity of soils in fields in terms of REV, it has been assumed tacitly that every REV is independent of each other. This means that every physical property given as an averaged value within each REV is independent. It is, however, often recognized that the differences in soil properties depend on their mutual distances (i.e., the closer the two arbitrarily chosen REVs, the lesser the differences in their properties). Taking account of the histories of soils in the sense of geographical, topographical, and geological influences on the properties of natural soils, it is not surprising that a soil has features similar to those of the surrounding

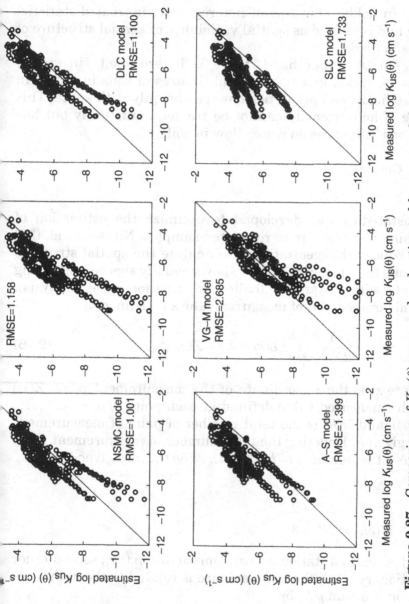

Figure 9.27 Comparison of $K_{us}(\theta)$ measured and estimated by means of the six models for sandy loam soils. The number of pairs of data is 247. (After Zhuang, J., Nakayama, K., Yu, G. R., and Miyazaki, T., *Soil Tillage Res.* 59:143–154 (2001). With permission.)

soils in fields. This type of heterogeneity is treated statistic-ally and is termed as spatial variability or spatial structure of soils in fields.

On the other hand, there is heterogeneity in strata, gravels, rocks, macropores, and fissures of soils in fields. In these fields, soil properties change abruptly with space. This type of heterogeneity cannot be treated statistically but has serious influences on water flow in soils.

B. Geostatistics

1. Semivariogram

Geostatistics was developed to optimize the estimation of mining reservoirs from restricted samples. Nielsen et al. (39) introduced this geostatistics to analyze the spatial structure of soils in fields. Figure 9.28 shows equally spaced measuring points with distance h (called *lag* in time-series analysis). Semivariance $\gamma(h)$ of measured values is defined by

$$\gamma(h) = \frac{1}{2N(h)} \sum_{i=1}^{N(h)} [Z(x_i + h) - Z(x_i)]^2 \qquad (9.46)$$

where x_i is the x-coordinate of the measurement point, $Z(x_i)$ is the measured value defined at each point x_i ($i = 1, 2, \ldots, N(h)$), and $N(h)$ is the total number of pairs of measurement for given h. Denoting the total number of measurement point by N^*, the numbers $N(h)$, $N(2h)$, $N(3h)$, ... are given by

$$N(h) = N^* - 1$$
$$N(2h) = N^* - 2$$
$$N(3h) = N^* - 3$$

and so on. In a stationary random condition (i.e., second-order stationary (40)), the semivariance is related to the autocorre-lation function $\rho(h)$ by

$$\gamma(h) = \sigma^2[1 - \rho(h)] \qquad (9.47)$$

where σ^2 is the variance of variable $Z(x_i)$. A typical semi-variogram and simultaneously determined coefficient of

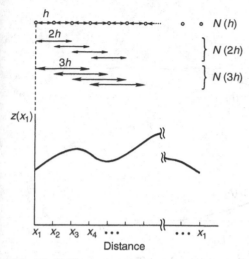

Figure 9.28 Equally spaced measuring points with lag h in the x-axis and distribution of measured values $Z(x_i)$.

autocorrelation as functions of lag h are given schematically in Figure 9.29, where point a designates the range of influence. In this example, any measured values $Z(x)$ are related more or less to other values $Z(x + h)$ when h is less than the range of influence a. When h exceeds this value, the semivariance $\gamma(h)$ reaches the sill C, which is identical with the variance σ^2 under a stationary random condition. It is concluded that the independence of REV in heterogeneous fields is guaranteed statistically only when each REV is separated more than the range of influence a.

2. Models of Semivariogram

Several types of models representing semivariogram have been proposed. The most common one is the spherical model (ideal model), given by

$$\gamma(h) = \begin{cases} C\left(\dfrac{3h}{2a} - \dfrac{h^3}{2a^2}\right) & \text{when } h \le a \\ C & \text{when } h \ge a \end{cases} \tag{9.48}$$

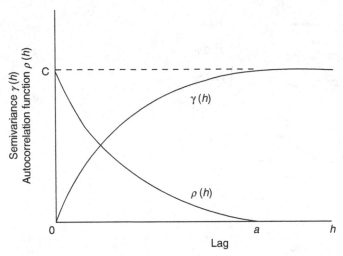

Figure 9.29   Typical semivariogram and coefficient of autocorrelation as functions of lag h.

where a is the range of influence and C is the value of sill. The linear model is given by

$$\gamma(h) = \begin{cases} C_0 + ph & \text{when } h \leq a \\ C & \text{when } h \geq a \end{cases} \qquad (9.49)$$

where C_0 is the nugget effect (Figure 9.30). When soil in a field has local heterogeneity as to clay lumps, rocks, and foreign materials, the value of semivariance with even a very small lag may be still large, due to the local heterogeneity. Hence, the nugget effect, C_0, increases with the local heterogeneity of soils.

3.   Spatial Structure in Fields

When a field is highly structured geostatistically, the range of influence may be large, while when a field is less structured geostatistically, the range of influence may be small. If there is no spatial structure at all in a given field, the range of influence is zero and any soil samples are regarded to be independent of each other.

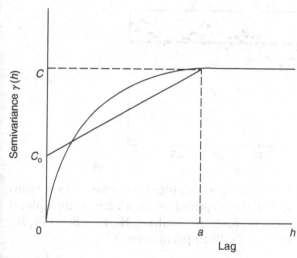

Figure 9.30 Spherical model (ideal model) and linear model for semivariogram.

Empirically, the range of structure is some meters in a less structured field and some tens of meters in a highly structured field. The ranges of influence depend on the properties of soil in question and on the size of the REV in a given field. This means that even if no spatial structure is recognized with respect to a given REV, there exists room to investigate spatial structures with respect to different sizes of REV. In addition, even if no spatial structure is observed with respect to a particular property in a given field, it remains possible to locate spatial structures of other properties in the same field.

Figure 9.31 shows an example of a semivariogram of mechanical impedance in a transection of silty loam soil at a depth of 30 to 45 cm in an upland field measured by Selim et al. (41). When the linear model (Equation (9.49)) was applied to the semivariogram, the range of influence was estimated to be 8.63 $\pm$ 1.30 m, the nugget effect was estimated to be 0.76 $\pm$ 0.09 MPa^2, and the sill was estimated to be 1.40 $\pm$ 0.03 MPa^2.

Figure 9.32 shows another sample semivariogram of bulk densities and volumetric water contents of surface soils in a

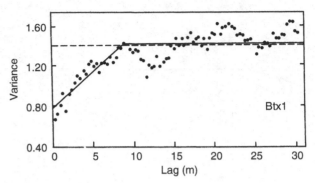

Figure 9.31 Semivariogram of mechanical impedance in a transection of silty loam soil at the depth of 30 to 45 cm in an upland field. (After Selim, H. M., Davidoff, B., Fluhler, H., and Schulin, R., *Soil Sci. 144*(6):442–452 (1987). With permission.)

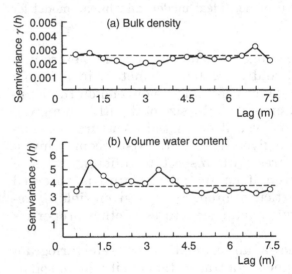

Figure 9.32 Semivariogram of (a) bulk densities and (b) volumetric water contents of surface soils in a paddy field. (After Haraguchi, N., *Trans. Jpn. Soc. Irrig. Drain. Reclam. Eng. 150*:27–35 (1990). With permission.)

paddy field, where no spatial structure was recognized (42). The range of influence is therefore zero in this field.

There are two details of practical significance to consider when investigating spatial structures in fields. First, once a spatial structure is identified in a field with a given REV, to obtain independent data the measurement or sampling must be conducted at locations separated from each other by more than the range of influence. Data taken within the range of influence are not independent of each other. Second, when the spatial structure is known, more precise estimations of unknown values of soil properties in a given field are available through a special statistical procedure, *kriging*.

4. Geostatistical Estimation

If a particular property of soil is known at locations 1 through 5 in a field as illustrated in Figure 9.33, the unknown value at location A is traditionally given simply by averaging the adjacent data as

$$f_A = \frac{1}{5} \sum_{i=1}^{5} f_i \qquad (9.50)$$

where f_A is the value in question at the location designated by A and f_i denotes the known value at the locations denoted by $i = 1, \ldots, 5$.

A better estimation of the unknown value at location A is obtained by taking account of the spatial structure in the calculation (43). By allocating weights ω_i to each sample

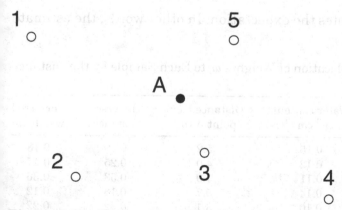

Figure 9.33 Estimation of the value at A by using known values at 1 through 5.

according to its distance from the point A, the estimator becomes

$$f_A = \sum_{i=1}^{5} \omega_i f_i \tag{9.51}$$

Table 9.5 gives an example where the volumetric water contents of soil samples at each location in Figure 9.33 are designated by distance from point A. By correcting the inverse distances so as to make the sum total to unity, the corrected weights ω_i are allocated to each sample.

The value of volumetric water content estimated by using Equation (9.50) is 0.124 cm^3 cm^{-3}, while that by using Equation (9.51) is 0.118 cm^3 cm^{-3}. The latter estimation is more reliable in this field. Using uranium estimation, Clark (43) demonstrated that the inverse distance estimation produces a (slightly) more accurate result than the arithmetic mean.

Kriging is a further generalized procedure in the determination of the weights ω_i. In kriging, the values of ω_I ($i = 1, 2, \ldots, n$) are determined by two conditions:

1. The expected value of the difference between true unknown value f_A and estimated value f_A^* is

$$E[f_A^* - f_A] = 0 \tag{9.52}$$

where E denotes the expectation. In other words, the estimate is unbiased.

Table 9.5 Allocation of Weights ω_i to Each Sample by the Distance from Point A

Sample number i	Water content (cm^3 cm^{-3})	Distance from point A (m)	Inverse distance	Corrected weight ω_i
1	0.15	5.2	0.19	0.13
2	0.12	4.0	0.25	0.17
3	0.11	1.9	0.53	0.36
4	0.14	5.7	0.18	0.12
5	0.10	3.1	0.32	0.22
			1.47	1.00

2. The variance

$$E\left[\{f_A^* - f_A\}^2\right]$$ (9.53)

is minimum.

The estimation of f_A^* in the case given in Figure 9.33 is defined by

$$f_A^* = \sum_{i=1}^{5} \omega_i f_i$$ (9.54)

where

$$\sum_{i=1}^{5} \omega_i = 1$$ (9.55)

In this procedure, the weights ω_i are termed the *kriging factors*. The detailed procedure for finding the kriging factors ω_i under these two conditions is beyond the scope of this book. Readers interested in kriging calculations are referred, for example, to the books by Clark (43) and Davis (44).

VI. MACROPORES

A. Types of Macropores

The heterogeneity of fields, formed by soil layers, surface crusts, cracks, gravels, rocks, and large pores due to small animals and plant roots, influences the flow of water in soils complexly. Differing from the spatial structure in fields, where gradual changes in soil properties in space predominate, this type of heterogeneity is accompanied by abrupt changes in soil properties in space, with evident boundaries against the soil matrix. The refraction of flow passing across these boundaries has been discussed in Chapter 3.

The effect of macropores on water flow was not studied comprehensively until Beven and Germann (45) published an epoch-making paper entitled "Macropores and Water Flow in Soils." Figure 9.34 illustrates exaggerated macropores in soils modified from the macropore model of Beven and Germann

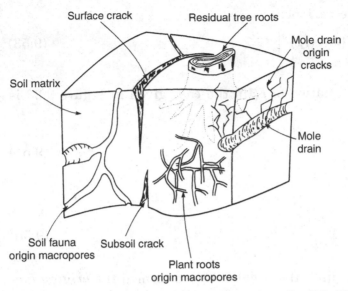

Surface crack   Residual tree roots

Mole drain
origin
cracks

Soil matrix

Mole
drain

Soil fauna   Subsoil crack
origin macropores
                    Plant roots
                  origin macropores

Figure 9.34 Exaggerated macropores in soils. (Modified from the macropore model given by Beven, K. and Germann, P., *Water Resour. Res. 18*:1311–1325 (1982). With permission.)

(45). The saturated water flow in these macropores, termed *bypassing flow*, is a type of preferential water flow (see Chapter 4). The macropores formed by large decayed plant roots occasionally develop similar root distribution configuration, resulting in bypassing flows in soils. Bypassing flows of water presumably appear in cracks in surface soils, in macropores formed by small animals, and in mole drains in subsurface soils provided that these bypasses are open to the atmosphere.

B. Buried Macropores

The most important criterion in the classification of macropores is whether the macropores are open in the atmosphere or they are buried under the soil surface.

The effects of buried macropores on water flow are not as simple. When the matric potential of water in a soil with macropores is less than zero (i.e., when the soil is under suction), water never moves out of the soil matrix and does not flow into macropores. Even when ponding water exists on the soil surface, the matric potential of subsurface water seldom exceeds zero, and water in such subsoil does not flow into the buried macropores.

Figure 9.35 shows an example of buried cracks formed in the clayey subsoil at a depth of 1 m below the surface in reclaimed land in Holland. The maximum width of the crack is about 10 cm. The groundwater level has been maintained below the cracks for a long time by the drainage facilities. It is supposed that the matric heads in the soil matrix around the cracks have been kept at a negative value, and hence no flow would have been generated within the cracks.

Buried macropores contribute to bypassing flows only when groundwater levels rise up to the macropores and the matric potential of the soil water exceeds zero. Furthermore, buried macropores contribute to subsoil drainage only when they have outlets into spaces in which the pressure is equal to that of the atmosphere. In this sense, buried cracks, buried mole drains, and buried macropores formed by plant roots and soil fauna, as shown in Figure 9.34, do not play a role in bypassing flow when the matric potential of the soil water is less than zero.

Figure 9.35 Typical buried cracks formed in clayey subsoil.

APPENDIX: THE b-VALUE ESTIMATION
OF CAMPBELL'S METHOD

The procedure of calculating the value of b in Equations (9.40) and (9.41) from Equations (9.38) and (9.39) is described in Campbell's textbook. It seems, however, convenient for us to summarize the procedure and to exemplify particular cases here.

The values of d_g and ρ_g are given by

$$d_g = \exp\left(\sum m_i \ln d_i\right)$$

$$\rho_g = \exp\left[\sum m_i (\ln d_i)^2 - \left(\sum m_i \ln d_i\right)^2\right]^{1/2}$$

where m_i is the mass fraction of textural class i and d_i is the arithmetic mean diameter of class i. According to Campbell, d_{clay} is 0.001 mm, d_{silt} is 0.026 mm, and d_{sand} is 1.025 mm.

For example, the Kunigami maji light clay (see Table 9.4) is composed of 34% clay, 34% silt, and 32% sand. By using the mass fractions given in the table and Campbell's d_i values, the parameters are determined as

$$d_g = \exp(0.34 \ln 0.001 + 0.34 \ln 0.026 + 0.32 \ln 1.025)$$
$$= 0.02783$$

$$\psi_{\text{es}} = -\frac{0.5}{\sqrt{d_g}} = -2.9972$$

$$\sigma_g = \exp[0.34(\ln 0.001)^2 + 0.34(\ln 0.026)^2$$
$$+ 0.32(\ln 1.025)^2 - (0.34 \ln 0.001 + 0.34 \ln 0.026$$
$$+ 0.32 \ln 1.025)^2]^{1/2}$$
$$= 16.69$$

and, eventually, the value of b is determined from Equations (9.38) to be 9.33.

REFERENCES

1. Cogels, O. G., Heterogeneity and representativity of sampling in the study of soil microstructure by the mercury intrusion method, *Agric. Water Manage. 6*:203–211 (1983).
2. Tokunaga, K. and Sato, T., The fundamental studies on the sampling method. II. On the variation in the distribution of the physical properties of soils found in large area farm land, *Trans. Jpn. Soc. Irrig. Drain. Reclam. Eng. 55*:1–8 (1975).
3. Sato, T. and Tokunaga, K., The fundamental studies on the sampling method. II. Analysis of sampled area variation of the distribution of the physical properties of soils found in large area farm land, *Trans. Jpn. Soc. Irrig. Drain. Reclam. Eng. 63*:1–7 (1976).
4. Rice, R. C. and Bowman, R. S., Effect of sample size on parameter estimations in solute transport experiments, *Soil Sci. 146*(2):108–112 (1988).
5. Lauren, J. G., Wagnent, R. J., Bouma, J., and Wosten, J. H. M., Variability of saturated hydraulic conductivity in a Glassaquic Hapludalf with macropores, *Soil Sci. 145*(1):20–28 (1988).
6. Inoue, H., Hasegawa, S., and Miyazaki, T., Lateral flow of water in an extremely cracked crop field, *Trans. Jpn. Soc. Irrig. Drain. Reclam. Eng. 134*:51–59 (1988).
7. Wood, E. F., Sivapalan, M., Beven, K., and Band, L., Effects of spatial variability and scale with implications to hydrologic modeling, *J. Hydrol. 102*:29–47 (1988).
8. Miller, E. E. and Miller, R. D., Physical theory for capillary flow phenomena, *J. Appl. Phys. 27*(4):324–332 (1956).
9. Childs, E. C. and Collis-George, N., The permeability of porous materials, *Proc. R. Soc. Lond. Ser. A 201*:392–405 (1950).
10. Campbell, G. S., *Soil Physics with BASIC: Transport Models for Soil-Plant Systems*, Elsevier, New York, pp. 50–53 (1985).
11. Scheidegger, A. E., *Physics of Flow Through Porous Media*, Macmillan, New York (1960).
12. Klute, A. and Wilkinson, G. E., Some tests of the similar media concept of capillary flow. I. Reduced capillary conductivity and moisture characteristic data, *Soil Sci. Soc. Am. Proc. 22*:278–281 (1958).
13. Wilkinson, G. E. and Klute, A., Some tests of the similar media concept of capillary flow. II. Flow systems data, *Soil Sci. Soc. Am. Proc. 23*:434–437 (1959).

14. Elrick, D. E., Scandrett, J. H., and Miller, E. E., Tests of capillary flow scaling, *Soil Sci. Soc. Am. Proc. 23*:329–332 (1959).

15. Tillotson, P. M. and Nielsen, D. R., Scale factors in soil science, *Soil Sci. Soc. Am. J. 48*(5):953–959 (1984).

16. Reichardt, K., Nielsen, D. R., and Biggar, J. W., Scaling of horizontal infiltration into homogeneous soils, *Soil Sci. Soc. Am. Proc. 36*:241–245 (1972).

17. Reichardt, K., Libardi, P. L., and Nielsen, D. R., Unsaturated hydraulic conductivity determination by a scaling technique, *Soil Sci. 120*(3):165–168 (1975).

18. Peck, A. J., Luxmoore, R. J., and Stolzy, J. L., Effects of spatial variability of soil hydraulic properties in water budget modeling, *Water Resour. Res. 13*(2):348–354 (1977).

19. Warrick, A. W., Mullen, G. J., and Nielsen, D. R., Scaling field-measured soil hydraulic properties using a similar media concept, *Water Resour. Res. 13*(2):355–362 (1977).

20. Warrick, A. W. and Amoozegar-Fard, A., Infiltration and drainage calculations using spatially scaled hydraulic properties, *Water Resour. Res. 15*(5):1116–1120 (1979).

21. Sharma, M. L. and Luxmoore, R. J., Soil spatial variability and its consequences on simulated water balance, *Water Resour. Res. 15*(6):1567–1573 (1979).

22. Simmons, C. S., Nielsen, D. R., and Biggar, J. W., Scaling of field-measured soil–water properties. I. Methodology. II. Hydraulic conductivity and flux, *Hilgardia 47*(4):77–173 (1979).

23. Russo, D. and Bresler, E., Scaling soil hydraulic properties of a heterogeneous field, *Soil Sci. Soc. Am. J. 44*:681–684 (1980).

24. Hillel, D. and Elrick, D. E., *Scaling in Soil Physics: Principles and Applications*, SSSA Special Publication 25, Soil Science Society of America, Inc., Madison, WI (1990).

25. Miyazaki, T., Bulk density dependence of air entry suctions and saturated hydraulic conductivities of soils, *Soil Sci. 161*:484–490 (1996).

26. Campbell, G. S., *Soil Physics with Basic: Transport Models for Soil–Plant Systems*, Elsevier, Amsterdam (1985).

27. Campbell, G. S. and Shiozawa, S., Prediction of hydraulic properties of soils using particle-size distribution and bulk density data, in *Indirect Methods for Estimating the Hydraulic Properties of Unsaturated Soils*, M. Th. van Genuchten, F. J. Leij, and L. J. Lund, Eds., U.S. Salinity Laboratory & U.C. Riverside, California, pp. 317–328 (1992).

28. Scheidegger, A. E., *The Physics of Flow Through Porous Media*, 3rd edn, University of Toronto Press, Toronto, pp. 137–144 (1974).

29. Cassel, D. K., Spatial and temporal variability of soil physical properties following tillage of Norfolk loamy sand. *Soil Sci. Soc. Am. J.* 47:196–201 (1983).

30. Zhuang, J., Yu, G. R., Miyazaki, T., and Nakayama, K., Modeling effects of compaction on soil hydraulic properties: an approach of non-similar media concept, *Adv. Geoecol.* 32:144–153 (2000).

31. Zhuang, J., Nakayama, K., Yu, G. R., and Miyazaki, T., Scaling of saturated hydraulic conductivity: a comparison of models, *Soil Sci.* 165:718–727 (2000).

32. Miyazaki, T. and Nishimura, T., Scaling of soils by using nonsimilar porous media concept. *Trans. Jpn. Soc. Irrig. Drain. Reclam. Eng.* 174:41–48 (1994).

33. Zhuang, J., Jin, Y., and Miyazaki, T., Estimating water retention characteristic from soil particle-size distribution using a non-similar media concept, *Soil Sci.* 166:308–321 (2001).

34. Zhuang, J., Nakayama, K., Yu, G. R., and Miyazaki, T., Predicting unsaturated hydraulic conductivity of soil based on some basic soil properties, *Soil Tillage Res.* 59:143–154 (2001).

35. Brooks, R. H. and Corey, A. T., Hydraulic properties of porous media. *Hydrology Paper 3*, Colorado State University, Fort Collins, CO (1964).

36. Poulsen T. G., Moldrup, P., and Jacobsen, O. H., One-parameter models for unsaturated hydraulic conductivity, *Soil Sci.* 163:425–435 (1998).

37. Alexander, L. and Skaggs, R. W., Predicting unsaturated hydraulic conductivity from soil water characteristic. *Trans. ASAE* 29:176–184 (1986).

38. van Genuchten, Mh. T. and Leij, F. J., On estimating the hydraulic properties of unsaturated soils, in *Indirect Methods for Estimating the Hydraulic Properties of Unsaturated Soils,* M. Th. van Genuchten, F. J. Leij, and L. J. Lund, Eds., U. S. Salinity Laboratory & U. C. Riverside, California, pp. 1–14 (1992).

39. Nielsen, D. R., Bigger, J. W., and Erh, K. T., Spatial variability of field-measured soil-water properties, *Hilgardia* 42(7):215–259 (1973).

40. Warrick, A. W. and Nielsen, D. R., Spatial variability of soil physical properties in the field, in *Applications of Soil*

Physics, D. Hillel, Ed., Academic Press, New York, pp. 338–343 (1980).

41. Selim, H. M., Davidoff, B., Fluhler, H., and Schulin, R., Variability of *in situ* measured mechanical impedance for a fragipan soil, *Soil Sci. 144*(6):442–452 (1987).

42. Haraguchi, N., On the change in the spatial pattern and spatial structure of soil moisture and bulk density caused by soil management, *Trans. Jpn. Soc. Irrig. Drain. Reclam. Eng. 150*:27–35 (1990).

43. Clark, I., *Practical Geostatistics,* Applied Science Publishers, London, pp. 102–106 (1979).

44. Davis, J. C., *Statistics and Data Analysis in Geology,* Wiley, New York, pp. 381–389 (1973).

45. Beven, K. and Germann, P., Macropores and water flow in soils, *Water Resour. Res. 18*:1311–1325 (1982).

Index

Printed in the United States
by Baker & Taylor Publisher Services